SMOOTHINGS OF PIECEWISE LINEAR MANIFOLDS

BY

MORRIS W. HIRSCH

AND

BARRY MAZUR

ANNALS OF MATHEMATICS STUDIES
PRINCETON UNIVERSITY PRESS

Annals of Mathematics Studies

Number 80

SMOOTHINGS OF PIECEWISE LINEAR MANIFOLDS

BY

MORRIS W. HIRSCH

AND

BARRY MAZUR

PRINCETON UNIVERSITY PRESS

AND

UNIVERSITY OF TOKYO PRESS

PRINCETON, NEW JERSEY

1974

Published in Japan exclusively by
University of Tokyo Press;
in other parts of the world by
Princeton University Press

Printed in the United States of America

Library of Congress Cataloging in Publication Data
will be found on the last printed page of this book.

PREFACE

The triangulation theorems of Whitehead [12] show that to every smooth manifold V there is associated a PL manifold M and a homeomorphism $f: M \to V$, having particularly nice local properties. Moreover M and f are unique up to very strong equivalence relations. When M is triangulated, f is called a smooth triangulation of V.

Whitehead's results suggest the inverse problem: is a given combinatorial manifold M the domain of a smooth triangulation of some smooth manifold V? This is possible if and only if M has a *smoothing*, i.e., a differential structure which is compatible with its piecewise linear structure.

Smoothing theory is concerned with the problem of finding and classifying smoothings of PL manifolds. The equivalence relation we use for the classification is not the obvious one of diffeomorphisms but the stronger relation of concordance: two smoothings of M are *concordant* if they extend to a smoothing of $M \times I$. This relation has nicer functorial properties than diffeomorphism. The diffeomorphism classification of smoothings is still unknown. It turns out that concordance is the same as the relation of isotopy: two smoothings α, β of M are *isotopic* if there is a diffeomorphism $M_\alpha \to M_\beta$ which is PD isotopic to the identity as a map $M \to M_\beta$.

This book is divided into two parts. Part I is devoted to proving that every smoothing of $M \times I$ is isotopic to a product smoothing. This is a very useful stability result; it implies for example that smoothings of $M \times \mathbf{R}^k$ correspond bijectively to smoothings of M, modulo concordance. (In Part II this is extended to smoothings of vector bundles over M.) Another consequence is that concordance and isotopy are the same equivalence relation. The methods used in Part I are entirely elementary.

Part II builds on the results of Part I to classify concordance classes of smoothings of M. For each PL manifold M, there is a "PL tangent bundle" t_M over M. Such bundles are classified by homotopy classes of maps $M \to BPL$ where PL is a certain semisimplicial group. There is a natural map $BO \to BPL$. If M is smoothable then t_M has a compatible vector bundle structure which we call a linearization; and isotopic smoothings define equivalent linearizations.

The main result of Part II is that *smoothings of* M *are completely classified by linearizations of* t_M. This basic fact can be rephrased in terms of classifying maps: M is smoothable if and only if the classifying map $f_M : M \to BPL$ of t_M lifts to a map $M \to BO$; and every homotopy class of liftings can be realized by a unique concordance class of smoothings.

If M is smoothable, it turns out that concordance classes of smoothings also correspond to homotopy classes of maps $M \to PL/O$, where PL/O is the fibre of $BO \to BPL$. The set $[M, PL/O]$ of such homotopy classes has a natural abelian group structure. See in this regard [8].

The classification of smoothings has been a model for several other classification theories. For example, Kirby and Siebenmann [2,13] have shown that the problem of putting a PL structure on a topological manifold M is equivalent to putting a PL structure on the "topological tangent bundle" of M – except perhaps in dimension 4. Sullivan [3,4], Browder [1] and Novikov [9] have analogous (but more complicated) results for the problem of classifying manifolds of a given homotopy type.

The present study is entirely theoretical in that no computations are carried out. The task of actually calculating the set of concordance classes of smoothings of a given manifold is still formidable. As a consequence of Part II, however, it "reduces" to standard problems in homotopy theory. Some results of this type are contained in the references.

REFERENCES

[1]. Browder, W., Manifolds and homotopy theory, in *Manifolds – Amsterdam 1970*. Lecture notes in mathematics 197, Springer-Verlag (Heidelberg) 1971.

[2]. Kirby, R., and Siebenmann, L., Some theorems on topological manifolds, *ibid.*

[3]. Sullivan, D., Geometric periodicity and the invariants of manifolds, *ibid.*

[4]. ————, Triangulating and smoothing homotopy equivalences. Mimeographed notes, Princeton University, 1967.

[5]. Munkres, J., Concordance of differentiable structures – two approaches. Michigan Math. Journal 14 (1967), 183-191.

[6]. ————, Obstructions to imposing differentiable structures. Illinois J. of Math. 18 (1964), 361-376.

[7]. ————, Higher obstructions to smoothing. Topology 4 (1965), 27-45.

[8]. ————, Concordance inertia groups. Advances in Math. 4 (1970), 224-235.

[9]. Novikov, S., Homotopy equivalent smooth manifolds I. Izv. Akad. Nauk. SSSR. Ser. Mat. 28 (1964), 365-474 = Amer. Math. Soc. Translations series 2, vol. 48, 271-396.

[10]. Scharlemann, M., and Siebenmann, L., The Hauptvermutung for smooth singular homeomorphisms, to appear.

[11]. Wall, C., an extension of results of Novikov and Browder. Amer. J. of Math. 88 (1966), 20-32.

[12]. Whitehead, J., On C^1 complexes. Annals of Math. 41 (1940), 809-832.

[13]. Kirby, R., and Siebenmann, L., On the triangulation of manifolds and the Hauptvermutung. Bull. Amer. Math. Soc. 75 (1969), 742-749.

CONTENTS

Smoothings of
Piecewise Linear Manifolds

SMOOTHINGS OF PIECEWISE LINEAR MANIFOLDS I: PRODUCTS

Morris W. Hirsch

§1. *Introduction*

The object of this paper is to give an elementary proof of the fundamental theorem of the theory of smoothings of PL manifolds:

MAIN THEOREM. *Let* α *be a smoothing of* $M \times I$, *inducing the smoothing* β *on* $M \times 0$; *let* $A \subset M$ *be a closed set such that* α *is a product smoothing in a neighborhood of* $A \times I$. *Given* $\varepsilon : M \times I \to R_+$ *there is a diffeomorphism from* $(M \times I)_\alpha$ *to the product smoothing* $M_\beta \times I$ *which is* PD ε*-isotopic to the identity rel* $M \times 0 \cup A \times I$.

A more elaborate theorem is stated in 4.1.

Versions of this theorem have been proved by Munkres [14] and independently, the author. Both of these proofs were based on Munkres' ingenious obstruction theory. The proof presented in this paper is independent of obstruction theory, and will be used in subsequent articles as the basis for the classification of smoothings; the classification has an obstruction theory as a corollary.

The origin of the main theorem was the announcement by R. Thom [20] in 1958 that every smoothing of $M \times I$ is diffeomorphic to product. Munkres' obstruction theory [12] has Thom's theorem as an immediate consequence.

A theorem weaker than the "$M \times I$ theorem" is the "$M \times \mathbf{R}$ theorem" 7.7:

Every smoothing $(M \times \mathbf{R})_\alpha$ *is isotopic to a product* $M_\beta \times \mathbf{R}$ *and* β *is unique up to concordance.* This was proved independently by Mazur-

Poenaru [11] and the author [3]. Using the main theorem, it follows *that* β *is in fact unique up to isotopy.*

Therefore: for every $n \in Z_+$, isotopy classes of smoothings of M correspond bijectively to isotopy classes of smoothings of $M \times R^n$. This is the first step in the isotopy classification of smoothings.

§2. *Structure of the paper*

In essence our method is that of Cairns [2] (see also [23]) who used fields of transverse planes to smooth submanifolds of Euclidean space.

If α is a smoothing of $M \times I$, we would like to simultaneously smooth all the level surfaces $M \times t$ of the projection $\pi_2 : (M \times I)_\alpha \to I$ by a small isotopy. This can be done provided there exists a vector field which is everywhere transverse to the level surfaces. By uniqueness of collaring we may assume that α is a product near $M \times 0$ and $M \times 1$. A small PL isotopy allows us to assume that $(M \times I)_\alpha$ has a smooth triangulation such that π_2 is simplicially affine and *separates vertices* in $M \times (I - \partial I)$. This ensures that every point other than a vertex has a tangent vector transverse to nearby level surfaces.

The most delicate part of the proof involves a close examination of the behavior of π_2 in the star of a vertex v. Everything is essentially determined by the level surfaces of π_2 restricted to the boundary of a small smooth ball B around v. If $\pi_2(v) = y_0$ then $M \times y_0 \cap \partial B$ is a PL sphere whose dimension is $m - 1 = \dim M - 1$, and its bounds a PL m-ball D in ∂B.

We can make D a smooth submanifold of B. By applying the main theorem one dimension lower as an induction hypothesis, we can assume D is a smooth m-disk. We then make D coincide with the northern hemisphere of ∂B by a diffeotopy.

Let us identify B with the unit ball $D^{m+1} \subset R^{m+1}$. By uniqueness of collaring of S^{m-1} in $\partial D^{m+1} = S^m$, we can assume the map π_2 agrees with the projection $\pi_{m+1} : R^{m+1} \to R$ on a neighborhood E of

S^{m-1} in S^m. Some careful smooth triangulations and the Alexander trick permit us to assume that π_2 and π_{m+1} agree on the cone on E from 0. One more simple radial isotopy makes π_2 and π_{m+1} agree in a neighborhood of 0. Since $\pi_{m+1} : D^{m+1} \to R$ is a smooth map, it follows that we have isotoped the smoothing α until the vertex $v \in (M \times I)_\alpha$ has a tangent vector transverse to the level surfaces near α.

The delicate part of the above argument lies in making sure that after each isotopy, we still have vectors transverse to the level surfaces at all points of the star of v other than v.

By a partition of unity we can find a globally defined vector field X in $(M \times I)_\alpha$ which is transverse to the level surfaces $M \times t$. By convolution we approximate π_2 by a smooth map $f : (M \times I)_\alpha \to I$ whose level surfaces are also transverse to X. We then push each $M \times t$ onto $f^{-1}(t)$ along integral curves of X. Finally if M is compact, we push each arc $x \times I$ onto the integral curve of X that starts at $x \times 0$.

If M is not compact, the last step is not always possible, for there is no guarantee that every trajectory of X passes from $M \times 0$ to $M \times 1$. We are forced to prove the theorem in stages: first the compact case, then a globalization to the non-compact case.

The globalization is not trivial because PD homeomorphisms do not form a category: in general they can be neither composed nor inverted. However, they "almost" form a category: a PD homeomorphism can be approximated by a PL homeomorphism. By using a refined version of J. H. C. Whitehead's theory of smooth triangulations, we obtain the globalization. It is remarkable that no change is required in Whitehead's proofs of thirty years ago; it is only necessary to examine them closely and extract their full strength.

"Relative" and "absolute"

In many statements of results to be proved there appears a manifold M and two closed subsets $K \subset N \subset M$. Their significance is the following. We want to improve some object (e.g., a map or a smoothing) in some way, in a neighborhood of K, without changing the object outside of N. A theorem that such improvement is possible we call *strongly relative*. If the theorem is in such a form that $K = N$, we call it *relative*, while if $K = M = N$ it is *absolute*.

Strongly relative theorems are the most useful although relative theorems often suffice to produce interesting theories. Absolute theorems are not as useful; they produce interesting theorems, but not theories.

To prove a strongly relative theorem, it usually is enough to prove it in a "local" form, where e.g., N is compact, or lies in a coordinate chart. Many results below follow this pattern. The globalization is ordinarily quite easy; occasionally we leave it to the reader.

Most, if not all, examples of strongly relative theorems can be placed in the context of sheaf theory: the "improved object" is a section of a sheaf J, and the strongly relative theorem says simply that every section of J over a closed subset of the base extends to a global section. We have, however, resisted the temptation to introduce sheaves. Nor has any attempt been made to formalize the notions of relative, strongly relative, etc.; rather they are used informally in discussing the logical pattern of the paper.

The main theorem, even in local form, seems to resist a direct proof. We first prove a local relative theorem, 4.2, which we refer to as the simplified form of the main theorem; it involves a weaker approximation. Then a local strongly relative theorem is proved, and finally an easy globalization yields the main theorem.

Contents of sections

Section 3 contains basic definitions and a brief discussion of diffeomorphism, isotopy and concordance. Section 4 contains statements of the main theorem (4.1) and its simplified form (4.2). Regular maps are studied in Sections 5 and 6. In Section 7 we digress from the proof of the main theorem to apply the preceding material to smoothings of submanifolds; some of these applications are used in Section 10. It is shown that smoothability depends only on PL tangential homotopy type. The proof of the simplified main theorem is completed in Sections 8 and 9; Sections 10 and 11 finish the proof of the main theorem itself. In Section 12 we study smoothings of $M \times \mathbf{R}^n$. The last section is an appendix in which we have collected various facts about PL and PD homeomorphisms; in particular the refined forms of Whitehead's smooth triangulation theorems are discussed.

§3. *Basic definition*

Differential structures

An *atlas* Φ on a topological space M is a set $\Phi = \{(\phi_i, U_i)\}_{i \epsilon \Lambda}$ of *charts* (ϕ_i, U_i), where $\{U_i\}_{i \epsilon \Lambda}$ is an open cover of M and each ϕ_i is a homeomorphism from U_i onto an open subset of $\mathbf{R}^n_+$. Here n is independent of i, and $\mathbf{R}^n_+$ is the closed halfspace

$$\mathbf{R}^n_+ = \{x \epsilon \mathbf{R}^n \mid x_n \geq 0\}$$

while $\mathbf{R}^n$ denotes n-dimensional Euclidean space.

A *manifold* is a metrizable space possessing an atlas.

A *differential structure* on a manifold M is an atlas $\Phi = \{(\phi_i, U_i)\}_{i \epsilon \Lambda}$ such that each map,

$$\phi_j \phi_i^{-1} : \phi_i(U_i \cap U_j) \to \phi_j(U_i \cap U_j)$$

is differentiable of class C^∞, or *smooth*, and maximal in this property. The pair (M, Φ) is called a *smooth manifold*. We shall frequently suppress notation for Φ.

If $f : N \to M$ is a homeomorphism from a space N onto an open subset of M, the *induced differential structure* $f^*\Phi$ on N is the differential structure containing the charts $(\phi_i f, f^{-1} U_i)$ for all $i \epsilon \Lambda$.

If (M_0, Φ_0) and (M_1, Φ_1) are smooth manifolds, a homeomorphism $f : M_0 \to M_1$ is a *diffeomorphism* from (M_0, Φ_0) to (M_1, Φ_1) if $f^*\Phi_1 = \Phi_0$.

Polyhedra

A *polyhedron* P is a subset of some Euclidean space R^n which can be expressed as a locally finite union of simplices. (Our simplices are always closed.) The simplices can always be chosen to form a simplicial complex C. We put $|C| = P$, and call C a *triangulation* of P.

If $P \subset R^n$ is a polyhedron, a map $f : P \to R^m$ is called *PL* if P has a triangulation C such that $f|\Delta : \Delta \to R^m$ is an affine map for every simplex $\Delta \in C$. We shall also say $f : C \to R^m$ is *simplicially affine*. If $f(P)$ is a polyhedron P' and $f : P \to P'$ is a proper map, then P and P' have triangulations C and C' making $f : C \to C'$ a simplicial map.

PL manifolds

Let M be a manifold. An atlas $\Phi = \{(\Phi_i, U_i)\}_{i \in \Lambda}$ for M is called PL provided each homeomorphism $\phi_j \phi_i^{-1} : \phi_i(U_i \cap U_j) \to \phi_j(U_i \cap U_j)$ is PL. A *PL manifold* is a pair (M, Φ) where Φ is a maximal PL atlas for M. We call Φ a *PL structure* for M.

A PL manifold (M, Φ) has a *triangulation* $T : P \to M$ where $P \subset R^q$ is a polyhedron, T is a homeomorphism, and each map $T^{-1} \phi_i^{-1} : \phi_i(U_i) \to P$ is PL.

Smooth triangulations

Let P be a polyhedron and N a smooth manifold. A homeomorphism $h : P \to N$ is called *PD* if P has a triangulation C such that $h|\Delta : \Delta \to N$ is a smooth map of rank dim Δ for each simplex $\Delta \in C$. We also say $h : C \to N$ is a smooth triangulation. It is known that the atlas on P given by the open stars of vertices of C makes P into a PL manifold.

If M is a PL manifold, a homeomorphism $h : M \to N$ is called PD if there exists a triangulation $T : P \to M$ such that the composition $T : P \to N$ is a PD homeomorphism.

If X and Y are spaces, an *isotopy* $f_t : X \to Y$ is a homotopy $f_t (0 \leq t \leq 1)$ such that the map $F : X \times I \to Y \times I$, $(x, t) : \mapsto (f_t(x), t)$ is a homeomorphism. If X and Y are polyhedra or PL manifolds and F is

PL, we call f_t a PL isotopy. If $f_t(x) = f_0(x)$ for all $t \in I$ and all $x \in A \subset X$, we call f_t a *rel A isotopy*. If M is a PL manifold or a polyhedra and N a smooth manifold, an isotopy $f_t : M \to N$ is called *PD* if the map $F : M \times I \to N \times I$ is a PD homeomorphism. (This is not well defined if $\partial N \neq \emptyset$, since in that case the natural "differential structure" on $N \times I$ has "corners," which we prefer to avoid. We require that the extension of F to $F' : M \times \mathbf{R} \to N \times \mathbf{R}$ be a PD homeomorphism, where $F'(x, t) = (f_1(x), t)$ for $t \geq 1$ and $F'(x, t) = (f(x), t)$ for $t \leq 0$). Let P be a polyhedron, N a smooth manifold, and $f_t : P \to N$ an isotopy. If C is a triangulation of P we call f_t *prismatic* in C if the map $(x, t) \mapsto (f_t(x), t)$ defines a smooth embedding $\Delta \times I \to N \times I$ of maximal rank, for each simplex $\Delta \in C$. Clearly a prismatic isotopy is PD.

J. H. C. Whitehead proved that every smooth manifold N admits a PD homeomorphism $M \to N$ and the PL manifold M is unique up to PL isomorphism. This theory is discussed in Section 13. The object of smoothing theory is to study the converse proposition from the point of view of M: classify the smooth manifolds N admitting a PD homeomorphism $M \to N$.

Smoothings of PL manifolds

A differential structure $\alpha = \{(\phi_i, U_i)\}_{i \in \Lambda}$ a PL manifold M is *compatible* with the PL structure $\Psi = \{(\psi_j, V_j)\}_{j \in \Sigma}$ if each map

$$\phi_i \psi_j^{-1} : \psi_j(U_i \cap V_j) \to \phi_i(U_i \cap V_j)$$

is a PD homeomorphism from the polyhedron $\psi_j(U_i \cap V_j) \subset \mathbf{R}^n$ onto the smooth submanifold $\phi_i(U_i \cap U_j) \subset \mathbf{R}^n$. A compatible differential structure on M is called a *smoothing of* M.

An equivalent definition is this: a differential structure α on a PL manifold M is compatible provided M has a triangulation $T : C \to M$ such that $T | \Delta : \Delta \to M_\alpha$ is a smooth map of rank dim Δ into the smooth manifold M_α, for each simplex $\Delta \in C$.

Concordance, isotopy and diffeomorphism

Let S(M) denote the set of smoothings of the PL manifold M. In S(M) there is the equivalence relation of diffeomorphism: we write $\alpha \approx \beta$ if the smooth manifolds M_α and M_β are diffeomorphic. We now consider some other relations on the set S(M).

For convenience assume $\partial M = \emptyset$. Then two smoothings α, β of M are *concordant* if there is a smoothing γ of $M \times I$ such that

$$\partial(M \times I)_\gamma = M_\alpha \times 0 \cup M_\beta \times 1 .$$

It is easy to see this is an equivalence relation. We call γ a *concordance* from α to β.

Let $K \subset M$ be any subset. We say α is *rel* K *isotopic* to β, written $\alpha \cong \beta$ rel K, if there exists a PD isotopy $f_t : M \to M_\beta$ such that:

$$f_0 = 1_M \text{ (the identity map of M)},$$

$$f_1 \text{ is a diffeomorphism } M_\alpha \to M_\beta, \text{ and}$$

$$f_t \mid K = 1_K \text{ for all } t \in I .$$

We call f_t a rel K isotopy from α to β_1 and write $f_t : \alpha \cong \beta$ rel K.

It is clear that the relation of rel K isotopy on S(M) is reflexive. It is also symmetric and transitive. To prove symmetry, suppose $f_t : \alpha \cong \beta$ rel K. Then $g_t : \beta \cong \alpha$ rel K, where $g_t = f_1^{-1} f_{1-t}$. (I am indebted to Derek White for this formula.) To prove transitivity, suppose

$$f_t : \alpha \cong \beta \text{ rel K and}$$

$$g_t : \beta \cong \gamma \text{ rel K}$$

Then $h_t : \alpha \cong \gamma$ rel K where

$$h_t : \begin{cases} g_{2t} & \text{if } 0 \leq t \leq \frac{1}{2} \\ g_1 f_{2t-1} & \text{if } \frac{1}{2} \leq t \leq 1 . \end{cases}$$

If M is metric and $d(f_t(x), x) < \varepsilon(x)$ for a map $\varepsilon : M \to R$ we call f_t an *ε-isotopy* and write $\alpha \cong_\varepsilon \beta$ if $f_1^* \beta = \alpha$. The relation "$\alpha \cong_\varepsilon \beta$ rel K for every $\varepsilon : M \to R_+$" is an equivalence relation.

It is clear that isotopy implies diffeomorphism. It is almost as obvious that isotopy implies concordance. To see this, suppose $f_t : \alpha \cong \beta$ and define $F : M \times I \to M_\beta \times I$ by $F(x,t) = (f_t(x), t)$. If $\partial M = \emptyset$, let $\gamma \in S(M \times I)$ be the smoothing induced from $\beta \times \iota$ by F, where $\iota \in S(I)$ is the standard smoothing, so that $M_\beta \times I = (M \times I)_\gamma$. Then γ is a concordance from β to α.

It was stated by Thom [21] and proved by Munkres [14] and the author (unpublished), that concordance implies isotopy. This is a corollary of the main theorem, whose essence is that *every concordance* $\gamma \in S(M \times I)$ *from* α *to* β *is isotopic* rel $M \times 0$ *to the product smoothing* $\alpha \times \iota$. Therefore if $h_t : \gamma \cong \alpha \times \iota$ rel $M \times 0$, then $h_t | M \times 1 : \beta \cong \alpha$. As relations on $S(M)$, we have:

$$\text{concordance} = \text{isotopy} \Longrightarrow \text{diffeomorphism} .$$

In proving the main theorem, however we may only use the obvious relations:

$$\text{concordance} \Longleftarrow \text{isotopy} \Longrightarrow \text{diffeomorphism} .$$

Concordance vs. diffeomorphism

To see that diffeomorphic smoothings need not be concordant, let M be a PL manifold having two nondiffeomorphic smoothings α and β. Let M', M'' be disjoint copies of M. Then $M'_\alpha \cup M''_\beta \approx M'_\beta \cup M''_\alpha$ but these two smoothings of $M' \cup M''$ are not diffeomorphic by any diffeomorphism homotopic to the identity. Therefore they are not isotopic. A connected example is given in Hirsch [3].

Another contrast between isotopy and diffeomorphism arises in considering extensions of smoothings. Let $M \subset V$ be a PL submanifold and α, β smoothings of M. If α extends to a smoothing of V and β is isotopic to α then β also extends. (This follows from a "PD isotopy extension theorem.'' Since we make no use of it in this paper, we leave the proof to the reader.)

If however α and β are merely diffeomorphic it is possible that α extends but not β. For example if $V = N \times I$ and $M = N \times \partial I$, choose $\gamma, \delta \in S(N)$ diffeomorphic but not concordant. Then the smoothing $\gamma \times 0 \cup \delta \times 1$ on $N \times \partial I$ does not extend over $N \times I$, but it is diffeomorphic to the smoothing $\gamma \times 0 \cup \gamma \times 1$ which obviously extends.

Such examples show there is no obstruction theory for classifying diffeomorphism classes of smoothings (see however Sullivan [20]). On the other hand, the goal of the concordance theory of smoothings is to show that functor which assigns to a smooth manifold M_α the set of concordance classes of smoothings of the PL manifold M is representable. There is a good obstruction theory for extending concordance classes of smoothings. This theory will be developed in later papers.

It is worth noting that the relation of diffeomorphism can hold between smoothings of *different* manifolds; but concordance and isotopy, make sense only for smoothings of the *same* manifold.

The relation of *prismatic isotopy* between smoothings has been studied by Kuiper [9].

We shall assume without further notice that smooth manifolds are metrized by the Riemannian metric induced from a smooth proper embedding in a Euclidean space, and that polyhedra or PL manifolds have a metric induced from a proper PL embedding in a Euclidean space. Such metrics are complete. Thus we can speak of ε-isotopies, etc., without specific mention of the metric.

§4. *Statement of the main theorem*

Throughout this section M is a PL manifold with empty boundary. A smoothing α of $M \times I$ is called *a product in a neighborhood of* $X \times I$ if $X \subset M$ has an open neighborhood $U \subset M$ such that $\alpha | (U \times I) = \beta \times \iota$ for some smoothing β of U. Equivalently the projection $\pi_1 : (U \times I)_\alpha \to U_\beta$ and $\pi_2 : (U \times I)_\alpha \to I_\iota$ are each smooth of maximal rank.

Let $K \subset N \subset M$ be closed sets. The boundary of N in M is denoted by $\mathrm{bd}N$ or $\mathrm{bd}_M N$. The diameter of N is denoted by $\mathrm{diam}\ N$.

We suppose M metrized, and we give $M \times I$ the metric

$$d((x_0, t_0), (x_1, t_1)) = \max\{d(x_0, x_1), |t_0 - t_1|\}.$$

4.1 MAIN THEOREM. *Suppose* $\alpha \in S(M \times I)$ *is a product smoothing in a neighborhood of* $(K \cap \mathrm{bd}N) \times I$. *For every* $\varepsilon : M \times I \to R_+$, α *is* ε*-isotopic rel* $(M \times 0) \cup (M - N) \times I$ *to a smoothing which is a product in a neighborhood of* $K \times I$.

We first prove the following *simplified form of the main theorem*. This is later globalized by technical devices to give the main theorem. The proof, in Section 9, involves an induction on $\dim M$.

4.2(m) THEOREM. *Let* $N \subset M$ *be compact, where* $\dim M = m$. Let $\varepsilon > 0$. *If* $\alpha \in S(M \times I)$ *is a product in a neighborhood of* $(\mathrm{bd}N) \times I$ *there exists* $\beta \in S(M \times I)$ *such that* $\alpha \cong_{d+\varepsilon} \beta$ rel $M \times 0 \cup (M - N) \times I$ *where* $d = \mathrm{diam}\ N$.

We prove 4.2 in Section 9, along with 9.1, by double induction on m. We shall use the following consequence of 4.2(m−1).

4.3(m) THEOREM (Munkres). *Let* B *be a smooth manifold which is* PD *homeomorphic to an* m*-simplex. Assume* 4.2(m−1). *Then* B *is diffeomorphic to the unit ball* $D^m \subset R^m$.

Proof. Let C be a triangulation of an m-simplex Δ and let $T : C \to B$ be a smooth triangulation.

Let $\sigma \in C$ be an m-simplex, and let $D_0 \subset \sigma - \partial\sigma$ be a smooth convex n-disk. Put $T(D_0) = D \subset B - \partial B$. Then D is a smooth submanifold of B diffeomorphic to D^m. It is easy to see that there is a PD homeomorphism

$$(\partial\Delta) \times I \to B - \mathrm{int}\ D.$$

By 4.3(m−1) with $M = \partial\Delta$, it follows that $B - \text{int } D \approx (\partial D) \times I$. Therefore B is diffeomorphic to D^m with a smooth collar $S^{m-1} \times I$ attached to $\partial D^m = S^{m-1}$ by a diffeomorphism. Hence $B \approx D^m$.

§5. *Regular maps*

Let $E \subset \mathbf{R}^n$ be a set and $f : E \to \mathbf{R}$ a map. For each $Y \in \mathbf{R}^n$, thought of as a vector, we define a map

$$L_Y f : E \to \mathbf{R} \cup \{+\infty, -\infty\}$$

by

$$L_Y f(a) = \text{Lim inf}_{(x,Z,t)\to(a,Y,0)}\, t^{-1}[f(x+tZ) - f(x)]$$

with the understanding that (x, Z, t) varies over the subset of $E \times \mathbf{R}^n \times \mathbf{R}$ such that x and $x + tZ$ are in E, and $t \neq 0$. If $L_Y f(a) > 0$, we say a is a *regular point* for f and that Y is *transverse to* f *at* a. This is devoted by $Y \pitchfork_a f$. An equivalent formulation is: there exist positive numbers ε, w such that

$$t^{-1}[f(x+tZ) - f(x)] \geq w$$

if $0 < |t| < \varepsilon$, $|x-a| < \varepsilon$, $|Z-Y| < \varepsilon$, and the left side of the inequality is defined. Any other point is *singular*.

The reader can easily supply proofs for the following fact.

5.1 LEMMA. *Let* $f, g : E \to \mathbf{R}$ *be maps defined on a subset* $E \subset \mathbf{R}^n$. *Then the following relations are true if the indicated sums and products exist (that is,* $(+\infty) + (-\infty)$ *and* $0 \cdot (\pm\infty)$ *are excluded)*

(a) $L_{Y+Z} f \geq L_Y f + L_Z f$

(b) $L_{cY} f = c L_Y f$ *if* $c > 0$

(c) $L_Y(f+g) \geq L_Y f + L_Y g$

(d) $L_Y(cf) = c L_Y f$ *if* $c \geq 0$

(e) $L_Y(f \cdot g) \geq f \cdot L_Y g + g \cdot L_Y f$ *if* f *and* g *are continuous and nonnegative*

(f) *If* f *is* C^1 *then* $L_Y f(a) = Df_a(Y)$

(g) $L_Y f(a) \leq \liminf_{(x,Z) \to (a,Y)} L_Z f(x)$.

Thus $L_Y f(a)$ *is lower semicontinuous in* (Y, a).

It follows from (g) that the set $\{(a,Y) \in E \times R^n | Y \pitchfork_a f\}$ is open in $E \times R^n$. From (a) and (b) the set $\{Y \in R^n | Y \pitchfork_a f\}$ is a convex open cone in R^n. From (c) and (d) we find that the set of $f : E \to R$ such that a is a regular point of f is a convex cone.

The following two lemmas are exercises in Taylor expansions.

5.2 LEMMA. *Let* $E, E' \subset R^n$ *be subsets and* $U, U' \subset R^n$ *neighborhoods of* E, E' *respectively; let* $\phi : (U, E) \to (U', E')$ *a* C^1 *diffeomorphism. Let* $f : E \to R$ *be a map. Let* $a \in E$, $Y \in R^n$, *and put* $a' = f(a)$, $Y' = D\phi_a(Y)$. *Then* $L_{Y'}(f\phi^{-1})(a') = L_Y f(a)$. *Therefore* $Y' \pitchfork_{a'} f\phi^{-1}$ *if and only if* $Y \pitchfork_a f$.

5.3 LEMMA. *Let* $a \in E \subset R^n$. *Let* $f : E \to R^n$ *be a map such that* $E = E_1 \cup \ldots \cup E_r$, *where* $a \in E_1 \cap \ldots \cap E_r$, *and* $f | E_i$ *extends to a* C^1 *map* f_i *defined on a neighborhood of* E_i. *Then for every* $Y \in R^n$

$$L_Y f(a) = \min\{Df_i(a)(Y) | i = 1,\ldots,r\} .$$

In particular

$$Y \pitchfork_a f \ \text{ if } \ Df_i(a)(Y) > 0 \ \text{ for } \ i = 1,\ldots,r .$$

The proof of the following technical result is complicated by the fact that the Lipschitz constant $\mathrm{Lip}(\phi h - \phi)$ does not necessarily approach 0 as $\mathrm{Lip}(h-1) \to 0$. See Section 13; and [7], Remark 1 of Section 1.

5.4 LEMMA. *Let* $E \subset R^n$ *be a set and* $\phi : E \to R$ *a map. Let* $F \subset E$ *be compact such that* ϕ *is regular on* F. *For every neighborhood* $U \subset E$ *of* F *there exists* $\delta > 0$ *such that if a map* $h : E \to E$ *satisfies*

(a) $Lip(h|U - 1_U) < \delta$ *and* $|h|U - 1_U| < \delta$, *then* $\phi h : E \to R$ *is regular on* F.

Proof. By compactness it suffices to prove the lemma for the case where F is a point x_0.

Let $Y_0 \in R^n$, $\varepsilon > 0$, $w > 0$ be such that

$$(1)\quad t^{-1}[\phi(x+tY) - \phi(x)] > w \ \text{ if } \ 0 < |t| < \varepsilon,\ |x - x_0| < \varepsilon,\ |Y - Y_0| < \varepsilon$$

and the left side of (1) is defined. If (a) is true we can write

$$(2)\quad h(x+tY) = z + tZ, \ \text{ with } \ |z - x| < \delta \ \text{ and } \ |Z - Y| < \delta Y$$

provided x and $x + tY$ are in U. If δ, $|x - x_0|$, $|Y - Y_0|$ and $|t|$ are small enough then (1) and (2) give

$$t^{-1}[\phi h(x+tY) - \phi h(x)] = t^{-1}[\phi(z+tZ) - \phi(z)] > w .$$

Hence $Y_0 \pitchfork \phi h$ at x_0. QED.

If $Y : E \to R^n$ is a vector field on $E \subset R^n$ and $f : E \to R$ is a map, we define a map

$$L_Y f : E \to R \cup \{-\infty + \infty\}$$

by $x \mapsto L_{Y(x)} f(x)$. If $L_Y f(x) > 0$ for all $x \in E$, we write $Y \pitchfork f$ or $Y \pitchfork_E f$ and say Y is *transverse* to f or f is Y-*regular*.

The usual proof by convolution that continuous functions can be approximated by smooth function also proves the following result.

5.5 THEOREM. *Let* $U \subset R^n$ *be an open set,* $f : U \to R$ *a continuous map, and* $K \subset U$ *a compact set. Let* $Y : K \to R^n$ *be a vector field. Given* $\varepsilon > 0$, *there exists a smooth map* $g : K \to R$ *such that*

$$L_Y g \geq (L_Y f)|K - \varepsilon$$

and

$$|g-(f|K)| < \varepsilon .$$

In particular, if $Y \pitchfork_K f$ *we can choose* g *so that* $Y \pitchfork_K g$.

Proof. A *convolution kernel* is a smooth nonnegative map $\lambda : R^n \to R$ such that $\int_{R^n} \lambda = 1$, and there exists $r(\lambda) > 0$ such that $\lambda(x) = 0$ if $|x| \geq r(\lambda)$. The *convolution*

$$f * \lambda : x \mapsto \int_{R^n} f(x-s)\lambda(s)\,ds$$

is defined on the set

$$U' = \{x \in U \mid d(x, R^n - u) < r(\lambda)\} .$$

It is well known that $f * \lambda$ is smooth, and as $r(\lambda) \to 0$, $f * \lambda \to f$ uniformly on compact sets. Suppose $r(\lambda)$ so small that $K \subset U'$. If $\varepsilon > 0$, by compactness there exists $\delta > 0$ such that

(1) $t^{-1}[f(x+tX)-f(x)] \geq L_{Y(a)}f(a) - \varepsilon$

where $x \in U$, $a \in K$, $X \in R^n$ are such that the left side of (1) is defined and $0 < |t| < \delta$, $|x-a| < \delta$, $|X - Y(a)| < \delta$. To estimate $L_{Y(a)}(f*\lambda)(a)$ we have $t^{-1}[(f*\lambda)(x+tZ)-(f*\lambda)(x)] =$

(2) $t^{-1} \int_{R^n} [f(x-s+tZ)-(x-s)]\lambda(s)\,ds$.

If $r(\lambda)$ and $|x-a|$ are sufficiently small then, since we may assume $|s| \leq r(\lambda)$, $|(x-s)-a| < \delta$.

Therefore (2) will be not less than $L_{Y(a)}f(a) - \varepsilon$ provided $r(\lambda)$ and $|x-a|$ are sufficiently small. It follows that $g = f * \lambda$ will fulfill the condition of the theorem if $r(\lambda)$ is sufficiently small.

We shall need an addendum to 5.5 in the special case where K is the *half-disk*

$$D^n_+ = D^n \cap R^N_+ = \{x \in R^n \mid x_n \geq 0 \text{ and } |x| \leq 1\} .$$

Observe that $D^n_+ \cap R^{n-1} = D^{n-1}$.

5.6 THEOREM. *Suppose that* $K = D^n_+, f|U \cap R^{n+1}$ *is constant, and* $Y \pitchfork_K f$, *in Theorem* 5.4. *Then* g *can be chosen to have the additional property that* $g|D^{n-1} = f|D^{n-1}$.

Proof. We may assume the constant value of $f|U \cap R^{n-1}$ to be 0.

First we replace f by a map $f_0 : U_0 \to R$ where $U_0 \subset U$ is a neighborhood of D^n_+ and f_0 is an extension of $f|U_0 \cap R^n_+$ such that:

(1) $f_0|(R^{n-1} \times t) \cap U_0$ is constant for all $t \leq 0$

(2) $L_Y f_0 \geq L_Y f$ on D^n_+.

We now construct f_0. Since $L_Y f$ is lower semicontinuous, there is a compact neighborhood V of D^n_+ in U so small that a lower bound

$$\alpha = \inf \{L_{Y(x)} f(x) \mid x \in V\} > 0$$

exists.

Next, observe that since $f|U \cap R^{n-1}$ is constant, and $L_Y f > 0$, it is impossible that $Y(x) \in R^{n-1}$ for $x \in R^{n-1} \cap U$. Therefore we may assume that $Y(x) = (Y_1(x), \ldots, Y_n(x))$ with $Y_n(x) > 0$ for $x \in R^{n-1} \cap U$. Let $\beta > 0$ be such that $Y_n(x) > \beta$ for all $x \in V \cap R^{n-1}$. Now set

$$f_0(x) = 2\alpha\, x_n\, \beta^{-1}$$

if $x \in V \cap R^n_-$, and $f_0(x) = f(x)$ for $x \in V \cap R^n_+$. We shall show that

(3) $L_{Y(x)} f_0(x) \geq \alpha$

for x in a neighborhood U_0 of D^n_+ in V. Clearly (3) holds if $x_n > 0$ or $x_n < 0$. Suppose then that $u \in V \cap R^{n-1}$ so that $u_{n-1} = 0$. Since $L_{Y(x)} f(x) \geq \alpha$ for all $x \in V$, given $\varepsilon > 0$ there exists $\delta > 0$ such that, for all x, $a \in V$.

(4) $0 < |t| < \delta$, $|x-a| < \delta$ and $|Z - Y(a)| < \delta$.

Choose $r > 0$ so small that $r < \delta$, and if $|x-u| \leq r$ and $|X-Y(u)| \leq r$, then

(5) $X_n > \beta$

and

(6) the interval $[x, x+rX]$ lies in V and in the ball of radius δ around u.

Now assume

(7) $u \in V \cap R^{n-1}$, $|x-u| < r$, $|X-Y(u)| < r$, and $0 < |t| < r$.

If $[x, x+tX] \subset R^n_-$ then

$$t^{-1}[f_0(x+tX)-f_0(x)] = 2\alpha\, X_n\, \beta^{-1} \geq 2\alpha ,$$

since $X_n > \beta$. If $[x, x+tX] \subset R^n_+$, then

$$t^{-1}[f_0(x+tX)-f_0(x)] > \alpha - \varepsilon .$$

Consider the case

$$x \in V \cap R^n_- \text{ and } x + tX \in V \cap R^n_+,\ 0 < t \leq r .$$

Then there exists $0 < s < t$ such that

$$y = x + sX \in V \cap R^{n-1} .$$

Hence

$$t^{-1}[f_0(x+tX)-f_0(x)] =$$

$$= (st^{-1})s^{-1}[f_0(x+sX)-f_0(x)] + (t-s)t^{-1}(t-s)^{-1}[f(y+(t-s)X-f(y)]$$

$$\geq st^{-1}\, 2\alpha + (t-s)t^{-1}(\alpha-\varepsilon) < \alpha - \varepsilon .$$

This proves that (7) implies

$$t^{-1}[f_0(x+tX)-f_0(x)] \geq \alpha - \varepsilon .$$

It follows that $L_{Y(u)} f_0(u) \geq \alpha$.

Continuing the proof of 5.6, consider the convolution $f_0 * \lambda$ where the convolution kernel $\lambda : R^n \to R$ has the special form

$$\lambda(x) = \mu(x_1)\ldots\mu(x_n)$$

where $\mu : R \mapsto R$ is a convolution kernel. It is clear that if $t \leq -r(\mu)$, then $f * \lambda$ is constant on the slice $U_0 \cap (R^{n-1} xt)$. The function $h(x) = f * \lambda(x_1, \ldots, x_{n-1}, x_n - r(\mu))$ take a constant value c on $U_0 \cap R^{n-1}$. Put $g(x) = h(x) - c$ thus $g(x) = 0$ if $x_n = 0$. If $r(\mu)$ is sufficiently small, then g satisfies Theorem 5.6.

Regular maps on smooth manifolds

If $f : M \to R$ is a map of a smooth manifold M and $Y \in T_xM$ is a tangent vector to M at a point $x \in M$, we define $L_Y f(x) \in \{+\infty, -\infty\}$ as follows.

Let $\phi : U \to R^n$ be a coordinate chart around x; put $\phi(x) = y \in R^n$ and $D\phi_x(Y) = Z \in R^n$. Then define

$$L_Y f(x) = L_Z(f \circ \phi^{-1})(y) .$$

By 5.2 this definition is independent of the chart (ϕ, U). If $L_Y f(x) > 0$ then x is a *regular point* for f; and Y is *transverse* to f at x denoted by $Y \pitchfork_x f$. If every point of M is regular for f, then f is a *regular* map.

If Y is a vector field on a subset $K \subset M$ and $Y(x) \pitchfork_x f$ for every $x \in K$, we say Y is *transverse* to f, or f is Y *regular* and write $Y \pitchfork f$ or $Y \pitchfork_K f$.

5.7 LEMMA. *Let* M *be a smooth manifold and* $f : M \to R$ *a regular map. Then:*

(a) *There exists a vector field on* M *transverse to* f.

(b) *If* Y *is a vector field on* M *transverse to* f *and* Y *is smooth in a neighborhood of a closed set* $X \subset M$, *then for every* $\varepsilon : M \to R_+$ *there exists a smooth vector field* Z *transverse to* f *such that* $|Z - Y| < \varepsilon$ *and* $Z = Y$ *in a neighborhood of* X.

Proof. (a) Given $x \in M$, let Y be a vector field defined on a neighborhood of x such that $Y(x) \pitchfork_x f$. Since this mean $L_{Y(x)}f(x) > 0$ it follows that $Y \pitchfork_U f$ for some neighborhood U of x. Let $\{U_i\}_{i \epsilon \Lambda} = U$ be an open covering of M such that for each $i \epsilon \Lambda$ there is a vector field Y_i on U_i with $Y_i \pitchfork (f|U_i)$. Let $\{P_i\}_{i \epsilon \Lambda}$ be a partition of unity subordinate to U and define $Y(x) = \Sigma\, P_i(x) Y_i(x)$, summed over $\{i \epsilon \Lambda | P_i(x) > 0\}$. From 5.1(a,b) we have

$$L_{Y(x)} f(x) \geq \sum P_i(x)\, L_{Y_i(x)} f(x) > 0 \ .$$

Hence $Y \pitchfork f$. Part (b) follows from the approximation of continuous vector fields by smooth ones and 5.1(g).

The global approximation theorem for regular maps is the following strongly relative theorem.

5.8 THEOREM. *Let* Y *be a vector field on a smooth manifold* M, *transverse to* $f: M \to R$. *Let* $N \subset M$ *and* $K \subset N$ *be closed sets such that* f *is* C^r *in a neighborhood* $W \subset M$ *of* $K \cap \mathrm{bd}N$ *for some* $1 \leq r \leq \infty$. *Given* $\varepsilon: M \to R_+$, *there exists a map* $g: M \to R$ *such that:*

(a) g *is* Y*-regular*

(b) $L_Y g \geq L_Y f - \varepsilon$

(c) $g(x) = f(x)$ *for* $x \epsilon M - N$

(d) $|g - f| < \varepsilon$

(e) g *is* C^r *in a neighborhood of* K

(f) $f|C = g|C$ *for every component* C *of* ∂M *on which* f *is constant;*

(g) $g|Q$ *is smooth for every subset* $Q \subset M$ *on which* f *is smooth. In particular if* f *is smooth on each simplex of a given smooth triangulation of* M, *the same is true of* g.

Proof. Let $\{(U_i, \phi_i)\}_{i \epsilon \Lambda}$ be a C^∞ atlas on M having the following properties:

(1) $\phi_i(U_i) = R^n$ or R^n_+, according to whether $U_i \cap \partial M$ is empty or nonempty;

(2) the open covering $\{U_i\}_{i\epsilon\Lambda}$ is locally finite;

(3) the sets $\{\text{int } D_i\}_{i\epsilon\Lambda}$ cover M where $D_i = \phi_i^{-1} D^n$;

(4) if $D_i \cap K \cap \text{bd}N \neq \emptyset$, then $U_i \subset W$, and hence $f|U_i$ is C^r;

(5) if $D_i \cap (K-W) \neq \emptyset$ then $U_i \subset N$.

Choose a C^∞ partition of unity $\{P_i\}_{i\epsilon\Lambda}$ such that $\text{int } D_i = \{x \epsilon M | P_i(x) > 0\}$.

We may assume $0 < \varepsilon < L_Y f$.

Choose δ_i such that

$$0 < \delta_i < \min \left\{ L_Y f(x) - \frac{\varepsilon}{2}(x) \,|\, x \epsilon D_i \right\}.$$

Put $\Lambda' = \{i \epsilon \Lambda | D_i \cap (K-W) \neq \emptyset\}$.

By 5.5 and 5.6 there exists C^∞ maps $g_i : D_i \to R$ for $i \epsilon \Lambda'$, such that, for $x \epsilon D_i$

(6) $L_Y g_i(x) > L_Y f(x) - \delta_i < \varepsilon(x)/2$

(7) $|g_i(x) - f(x)| < \delta_i$

and

(8) $g_i(x) = f(x)$ if $x \epsilon D_i \cap C$ for each component C of ∂M on which f is constant.

If $i \epsilon \Lambda - \Lambda'$, define $g_i = f|U_i$.

Observe that $g_i = f_i$ on $M - N$, and g_i is C^r in a neighborhood of K. Moreover $g_i|Q \cap D_i$ is smooth if $f|Q$ is smooth. Define $g : M \to R$ by $x \mapsto \sum_{i\epsilon\Lambda(x)} P_i(x) g_i(x)$ where $\Lambda(x) = \{i \epsilon \Lambda | x \epsilon \text{ int } D_i\}$.

If the δ_i are sufficiently small, g will obviously satisfy all parts of the theorem except perhaps (a) and (b).

To obtain (b) and hence (a), observe that since the P_i are smooth, $\Sigma L_Y P_i = \Sigma DP_i(Y) = (\Sigma DP_i)(Y) = D(\Sigma P_i)Y = 0$. Therefore $(\Sigma L_Y P_i) f = 0$ and we have, summing over $\Lambda(x)$ for each point $x \epsilon M$:

$$\begin{aligned} L_Y g - L_Y f &= L_Y \Sigma P_i g_i - L_Y \Sigma P_i f - \Sigma L_Y P_i \\ &\geq \Sigma(L_Y P_i) g_i + \Sigma P_i L_Y g_i - L_Y \Sigma P_i f - (\Sigma L_Y P_i) f \\ &\geq \Sigma(L_Y P_i)(g_i - f) + \Sigma P_i (L_Y g_i - L_Y f) \\ &\geq -\Sigma |L_Y P_i| \, |g_i - f| + \Sigma P_i (L_Y g_i - L_Y f) \; . \end{aligned}$$

Put $A(x) = \sup \{ |DP_i Y(x)| : i \epsilon \Lambda(x) \}$. Then $L_Y P_i(x) \leq A(x)$, and hence from the above,

$$L_Y g(x) - L_Y f(x) \geq - A(x) \Sigma \delta_i + \frac{1}{2} \varepsilon(x) \; ;$$

we obtain (b) by choosing the δ_i so small that

$$A(x) \sum_{i \epsilon \Lambda(x)} \delta_i < \frac{1}{2} \varepsilon(x)$$

for all $x \epsilon M$. This completes the proof of Theorem 5.7.

§6. *Regular maps and isotopics*

Geometric properties of regular maps

6.1 LEMMA. *Let* $U \subset R^n$ *be an open set,* $f : U \to R$ *a map, and* $Y \epsilon R^n$ *a vector transverse to* f *at* $x_0 \epsilon U$. *Then the level surface* $f^{-1}f(x_0)$ *through* x_0 *is locally homeomorphic to* R^{n-1} *at* x_0.

Proof. Without loss of generality we may suppose that $x_0 = 0 \epsilon R^n, f(x_0) = 0 \epsilon R$, and that Y is the unit vector $(0, \dots, 0, 1)$. Since $Y \pitchfork_0 f$, there exists numbers $\varepsilon > 0$, $u > 0$ such that

(1) $t^{-1}[f(x+tY) - f(x)] > u$ if $0 < |t| < \varepsilon$ and $|x| < \varepsilon$.

By continuity of f there exists $0 < \delta < \varepsilon$ such that

(2) $|f(x)| < \frac{1}{2} \varepsilon u$ if $|x| \leq \delta$.

Put $R^n = R^{n-1} \times R$, and consider the restriction of f to the interval $I_y = y \times [-\varepsilon, \varepsilon]$ with $y \epsilon R^{n-1}$ and $|y| \leq \delta$.

By (1), $f : I_y \to \mathbf{R}$ is strictly increasing, and also $\varepsilon^{-1}[f(y,\varepsilon) - f(y,0)] > u$; hence $f(y,\varepsilon) > f(y,0) + u\varepsilon$. Therefore by (2)

(3) $f(y,\varepsilon) > 0$.

Similarly

(4) $f(y,-\varepsilon) < 0$.

Therefore there is a unique point $z \in I_y$ such that $f(z) = 0$. Therefore the set $f^{-1}(0) \cap I_v$ has only one element. Put $B^{n-1} = \{y \in \mathbf{R}^{n-1} : |y| \leq \delta\}$. The projection

$$\pi_1 : B^{n-1} \times [-\varepsilon, \varepsilon] \to B^{n-1}$$

maps the compact set

$$E = f^{-1}(0) \cap (B^{n-1} \times [-\varepsilon, \varepsilon])$$

bijectively onto B^{n-1}. By compactness,

$$\pi_1 \mid E : E \to B^{n-1}$$

is a homeomorphism. Since E is a neighborhood of 0 in $f^{-1}(0)$, the lemma is proved.

6.2 LEMMA. *Let* $U \subset \mathbf{R}^n$ *be open,* $f : U \to \mathbf{R}$ *a map, and* $Y : U \to \mathbf{R}^n$ *a vector field transverse to* f. *Then* f *is strictly increasing on each integral curve of* y.

Proof. Let $J \subset \mathbf{R}$ be an interval and $\lambda : J \to U$ an integral curve of Y. It suffices to prove that $f\lambda : J \to \mathbf{R}$ is locally increasing. Let $b \in J$ and set $x = \lambda(b) \in U$. Since $L_Y f > 0$, there exists $\varepsilon > 0$ such that:

(1) $f(x+tZ) - f(x) \geq te$ if $0 < |t| < \varepsilon$ and $|Z - Y(x)| < \varepsilon$.

Since $\lambda'(b) = Y(x)$, there exists $\delta > 0$ such that if $0 < t \leq \delta$, then $b + t \in J$ and

(2) $\lambda(b+t) = x + tZ$ with $|Z - Y(x)| < \varepsilon$.

Therefore if $0 < t < \delta$ we have

$$f\lambda(b+t) - f\lambda(b) = f(x+tZ) - f(x) \geq t\varepsilon > 0$$

by (1) and (2). The proof of 6.2 is complete.

Coregular maps

If $f: M \to \mathbf{R}$ is a regular map, there exists a smooth vector field Y transverse to f. We can approximate f by a smooth Y-regular map g. Then we can push each level surface of f onto the corresponding level surface of g along the integral curves of Y. In this may an isotopy h_t of M is obtained taking f into g. If α is the smoothing of M, then f is smooth in $h_1^*\alpha$. We proceed to elaborate this idea.

We summarize some well known fact about flows. Let X be a locally Lipschitz vector field on a smooth manifold M. The *X-flow* F_t is the collection of maximal integral curves $\lambda_x : t \mapsto F_t(x)$, for all $x \in M$. Each λ_x is defined on a connected subset $J_x \subset \mathbf{R}$ which contains 0, and $F_0(x) = x$. If $x \in M - \partial M$ then $0 \in \operatorname{int} J_x$. If $x \in \partial M$ and $X(x)$ is transverse to ∂M, then 0 is an endpoint of J_x, and J_x does not reduce to a single point.

If J_x contains a limit point y, then $\lambda_x(y) \in \partial M$. The set valued map $x \mapsto J_x$ is lower semicontinuous.

The map $(x,t) \to F_t(x)$ is continuous in (x,t) where it is defined and smooth if the field X is smooth. The equation $F_{s+t}(x) = F_s(F_t(x))$ is valid in the sense that if one side is defined, so is the other and they are equal. The derivative of λ_x at t is $X(\lambda_x(t))$.

Let M be a smooth manifold and $f, g : M \to \mathbf{R}$ maps. If X is a vector field on M, we say f and g are *X-coregular* provided:

(1) X is locally Lipschitz;

(2) both f and g are X-regular;

(3) if $x \in M - \partial M$ there is a number $\tau(x)$ such that $F_{\tau(x)}(x)$ is defined, where F_t is the X-flow, and $gF_{\tau(x)}(x) = f(x)$;

(4) $f|\partial M = g|\partial M$;

(5) the set $\{x \in M | f(x) \neq g(x)\}$ has compact closure.

Since f and g are increasing along integral curves of X the number $\tau(x)$ in (3) in unique. If we define $\tau(x) = 0$ for $x \in \partial M$, then the function $\tau : M \to \mathbf{R}$ is continuous. It is easy to see that the relation of X-coregularity is a equivalence relation on the set of X-regular maps $M \to \mathbf{R}$. If f and g are X-coregular we define the *X-homotopy relating* f *to* g to be the homotopy

$$h_t : M \to M,\ h_t(x) = F_{t\tau(x)}(x);\ 0 \leq t \leq 1 .$$

If each h_t is a homeomorphism we call h_t the X-*isotopy* relating f to g.

6.3 LEMMA. *For each* $t \in I$

(a) h_t *is injective,*

(b) h_t *is surjective if it is proper;*

(c) h_t *is bijective in case* $\{x \in M | f(x) \neq g(x)\}$ *has compact closure.*

Proof. (a) To prove h_t injective, fix $0 < t \leq 1$ and let $h_t(x) = h_t(y)$. Put

(1) $\tau(x) = u \leq \tau(y) = v$.

Apply F_{-tu} to $h_t(x)$ and $h_t(y)$, to get

(2) $x = F_{tv-tu}(y)$.

If $u = v$ then $x = y$, so suppose $u < v$. From (2) we see that x and y are on the same integral curve. Since $t(v-u) > 0$ and f increases along integral curves, from (2) we have $f(x) > f(y)$. The defining property of τ shows that

(3) $gF_u(x) > gF_v(y)$.

Put $z = F_{tu}(x) = F_{tv}(y)$. Then

(4) $F_u(x) = F_{(1-t)u}(z)$,

$F_v(y) = F_{(1-t)v}(z)$.

Since g increases along integral curves and

$$(1-t)v \geq (1-t)u \ ,$$

(4) gives

$$gF_v(y) \geq gF_u(x)$$

which contradicts (3). Therefore $x = y$, and so h_t is injective.

(b) To prove h_t surjective if it is proper, observe that $h_t(M)$ is then a closed subset of M. Since h_t is injective and $h_t|\partial M = 1$, $h_t(M)$ is an open set by invariance of domain. Therefore $h_t(M)$ is open and closed, and $h_t(\partial M) = \partial M$. It follows that $h_t(M) = M$.

(c) In this case $h_t = 1$ outside a compact set, so it is proper.

6.4 Lemma. *Let M be a smooth manifold with a smooth triangulation C. Let $f, g : M \to R$ be X-coregular maps, where X is a vector field transverse to ∂M. The X-homotopy $h_t : M \to M$ relating f to g is prismatic in C in each of the following cases:*

(a) *g and X are smooth, and f is smooth on each simplex of C.*

(b) *f, g and X are smooth on each simplex of C, and $X(x)$ is tangent to $\sigma \in C$ if $x \in \sigma - \partial\sigma$, for each $x \in M$, $\sigma \in C$.*

Proof. It suffices to prove that the map $\tau : M \to R$ is smooth on each simplex $\Delta \in C$, for then the defining formula

$$h_t(x) = F_{t\tau(x)}(x)$$

shows that h_t is prismatic. This is trivial in case (b). We proceed to (a).

To say that $f|\Delta$ is smooth means that for each $x_0 \in \Delta$ there is an open set $U \subset M$ containing x_0 and a smooth map $f_0 : U \to R$ such that $f_0|U \cap \Delta = f|U \cap \Delta$. If U is sufficiently small there is an interval $J \subset R$ containing 0 such that $F_t(x)$ is defined for all $(x, t) \in U \times J$. (We use here the assumption that X is transverse to ∂M.) Consider the smooth map

$$\Phi : U \times J \to R;\ (x,t) \mapsto gF_t(x) - f_0(x) \ .$$

We observe that $\frac{\partial\Phi}{\partial t}(x,0) = Dg\, X(x) > 0$ since $X \pitchfork g$. From the implicit function theorem we find a smooth map $\sigma : V \to R$ such that

$$g F_{\sigma(x)}(x) = f_0(x)$$

for some neighborhood V of x_0 in U. If $x \in V \cap \Delta$ then $f_0(x) = f(x)$ and hence $\sigma(x) = \tau(x)$. The existence of σ proves that $\tau|\Delta$ is smooth.

6.5 THEOREM. *Let* M *be a smooth manifold and* $N \subset M$ *a compact set. Let* $f : M \to R$ *be a regular map which is smooth in a neighborhood of* $\mathrm{bd}\, N$, *and constant on each component of* ∂M. *Given* $\varepsilon > 0$, *there exists a map* $g : M \to R$ *and an* ε*-isotopy* $h_t : M \to M$ *such that*

(a) g *is smooth of rank* 1 *in a neighborhood of* N

(b) $g = f$ *on* $(M - N) \cup \partial M$

(c) $gh_1 = f$

(d) h_t *is fixed on* $(M - N) \cup \partial N$

(e) *if* M *is smoothly triangulated so that* f *is smooth on each simplex, then* h_t *is a prismatic isotopy.*

Proof. By 5.6 there is a smooth vector field X on M such that f is X-regular. If C is a component of ∂M then $X|C$ is transverse to ∂M since $f|C$ is constant. Given $\delta > 0$, by 3.4 there exists a map $g : M \to R$ such that

(1) g is smooth in a neighborhood W of N;

(2) $g = f$ on $(M - N) \cup \partial M$;

(3) g is X-regular;

(4) $|g - f| < \delta$;

If δ is small enough, f and g will be X-coregular, and by 6.3 and 6.4, applied to $f|W$, $g|W$ and $X|W$, the X-homotopy relating f to g will satisfy the theorem.

We restate 6.5 in terms of isotopies of smoothings.

6.6 COROLLARY. *Let* α *be a smoothing of a* PL *manifold* M. *Let* $N \subset M$ *be a compact set and* $f: M_\alpha \to R$ *a regular map which is smooth in a neighborhood of* bd N *and constant on each component of* ∂M. *Suppose* M *has a smooth triangulation* C *in which* f *is simplicially smooth. Given* $\varepsilon > 0$ *there exists an* ε*-isotopy* $h_t: M \to M_\alpha$ *which is prismatic in* C *such that:*

(a) *if* $\beta = h_1^*\alpha$ *then* $f: M_\beta \to R$ *is smooth of rank* 1 *in a neighborhood of* N;

(b) h_t *is fixed on* $(M-N) \cup \partial M$.

The next result involves the notion of *isotopy of differential structures* on topological manifolds. This is defined like isotopy of smoothings of PL manifolds, leaving out all PL and PD considerations.

6.7 PROPOSITION. *Let* M *be a topological manifold without boundary* $N \subset M$ *a closed set, and* α *a differential structure of* $M \times I$ *which is a product in a neighborhood of* $(\text{bd}\, N) \times I$. *Suppose the projection* $\pi_2 (M \times I)_\alpha \to I$ *is smooth of rank* 1 *in a neighborhood of* $N \times I$. *If* N *is compact, then* α *is isotopic* rel $M \times 0 \cup (M-N) \times I$ *to a differential structure which is a product in a neighborhood of* $N \times I$ *by an isotopy which preserves* π_2. *Moreover if* M *is* PL *and* α *is compatible with the product* PL *structure on* $M \times I$, *the isotopy in* PD.

Proof. Endow $(M \times I)_\alpha$ with a Riemannian metric which is a product metric in a neighborhood $V \times I$ of $(\text{bd}\, N) \times I$ on which α is a product smoothing. For each $p \in (N \cup V) \times I$ let $X(p)$ be the vector tangent to $(M \times I)_\alpha$ at p, orthogonal to the smooth level surface $M \times \pi_2(p)$ of π_2, and such that $D\pi_2(X) = 1$. Thus $X(p) = \text{grad}\, _p\pi/_2 |\text{grad}\, _p\pi_2|^2$. If $p \in V \times I$ then $X(p)$ is the unit tangent to the curve $t \mapsto (x, t)$ passing through p, since the metric on $V \times I$ is a product.

Let $\gamma_x : J_x \to (M \times I)_\alpha$ be the maximal integral curve of the vector field X which starts at $(x, 0)$ for each $x \in N \cup V$. If $x \in V$ then $J_x = [0,1]$ and $\gamma_x(t) = (x, t)$. The condition $D\pi(X) = 1$ implies:

(1) $\pi\gamma_x(t) = t$ if $t \in J_x$.

Therefore $J_x = [0, a]$ or $[0, a\rangle$ with $a \leq 1$. If $x \in N - V$ then $\gamma_x(J_x) \subset (N-V) \times I$ because of uniqueness of integral curves. It follows that if $J_x = [0, a]$ then $a = 1$. Since $N \times I$ is compact, the case $J_x = [0, a\rangle$ is impossible, by (1), and the maximality of J_x. Hence $J_x = [0, 1]$ for all $x \in N \cup V$.

Let α induce the smoothing β on $M \times 0$. Define a diffeomorphism

$$G : (N \cup V)_\beta \times I \to [(N \cup V) \times I]_\alpha$$

by

$$G(x, t) = \gamma_x(t) .$$

Observe that G preserves π; we put

$$G(x, t) = (g_t(x), t) .$$

Define an isotopy

$$f_t : (N \cup V) \times I \to (N \cup V) \times I$$

by

$$f_t(x, s) = \begin{cases} (g_s(x), s) & \text{if } 0 \leq s \leq t \\ (g_t(x), s) & \text{if } t \leq s \leq 1 . \end{cases}$$

Extend f_t to be the identity on $(M-N) \times I$. Then f_1 is the identity and $f_0 = G$ on $(N \cup V) \times I$. The isotopy $h_t = f_{1-t}$ satisfies the proposition.

If N in 6.7 is not compact, the proof of 6.7 fails since the integral curves of the vector field X need not pass from $N \times 0$ to $N \times 1$. There is no obvious way globalizing 6.7, which is relative, but not strongly relative, in form; in fact the generalization to noncompact N is known only for the case where α is compatible with a product PL structure on $M \times I$; this follows from the main theorem.

The following theorem is also true for topological manifolds.

6.8 THEOREM. *Let* M *be a* PL *manifold without boundary,* $N \subset M$ *a closed set, and* α *a smoothing of* $M \times I$ *which is a product in a neighborhood of* $\text{bd}\, N \times I$. *Assume:*

(a) N *is compact and*

(b) $\pi_2 : (M \times I)_\alpha \to I$ *is regular on* $N \times I$.

Let $d = \text{diam}\, N$. *Given* $\varepsilon > 0$, α *is* $(d+\varepsilon)$*-isotopic* rel $M \times 0 \cup (M-N) \times I$ *to a smoothing of* $M \times I$ *which is a product in a neighborhood of* $N \times I$.

Proof. Combine 6.6 and 6.7.

Special situations

Suppose M is a smooth manifold endowed with a fixed smooth triangulation. Let $K \subset M$ be a subcomplex, and suppose the set of vertices of K is partitioned into two disjoint sets V_+ and V_-. Let K_+ and K_- denote the largest subcomplexes of K whose vertices are respectively in V_+ and V_-; then $|K_+|$ and $|K_-|$ are disjoint.

Every simplex σ of K is the join of its two faces

$$\sigma_+ = \sigma \cap |K_+| , \quad \sigma_- = \sigma \cap |K_-| .$$

Put $\hat{\sigma} = \sigma - (\sigma_+ \cup \sigma_-)$. Each point $x \in \hat{\sigma}$ lies on a unique interval $[x_-, x_+] \subset \sigma$ with $x_- \in \sigma_-$ and $x_+ \in \sigma_+$. We define a vector field X_σ on $\hat{\sigma}$ as follows. Write $x = (1-r)x_- + rx_+$, with $0 < r < 1$. Then $X_\sigma(x)$ is the tangent at $t = r$ to the curve $t \mapsto (1-t)x_- + t_x$, $0 \le t \le 1$. Clearly X_σ and X_τ agree on $\hat{\sigma} \cap \hat{\tau}$ for each pair σ, τ of simplices of K. Put $\hat{K} = |K| - (|K_-| \cup |K_+|)$ and define a vector field X on $\hat{K}$ in M by setting $X | \hat{\sigma} = X_\sigma$ for each simplex σ. Then X is continuous; in fact X is simplicially smooth and hence locally Lipschitz. In our application $\hat{K}$ will be an open set fibered by the integral curves of X.

6.9 LEMMA. *Let* $f: K \to R$ *be simplicially affine, and suppose* $f(x_-) < 0 < f(x_+)$ *for all* $x_- \in |K_-|$ *and* $x_+ \in |K_+|$. *Then* f *is X-regular on* $\hat{K}$.

Proof. Follows easily from 5.3.

If C is a simplicial complex and $p \in |C|$ the *star* $St(p) = St_C(p)$ is the union of all the simplices of C which contain p.

6.10 LEMMA. *Let* L *be a subcomplex of a smooth triangulation of* M; *let* $f: L \to R$ *be a simplicially affine. Suppose* $p \in |L|$ *is not a vertex. If* f *is not constant on vertices of* St(p) *then* p *is a regular point of* f.

Proof. We may suppose $f(p) = 0$. The lemma follows from 6.9 by separating the set of vertices v of St(p) into two nonempty sets according to whether $f(v) > 0$ of $f(v) < 0$.

The next lemma is similar to 6.5, where the X-isotopy relating X-coregular maps f and g was used. The difference lies in keeping the isotopy fixed outside a given neighborhood of a level surface. A general result is somewhat awkward, so we content ourselves with a very special situation. Let the subcomplex K of a smooth triangulation of M be as described above, along with vector field X and the set $\hat{K} \subset |K|$. Let $f: K \to R$ be a simplicially affine map as in 6.9. Suppose $\hat{K}$ is open in M, and put $M_0 = f^{-1}(0) \subset \hat{K}$.

6.11 LEMMA. *Let* $U \subset \hat{K}$ *be a neighborhood of* M_0. *Let* $A \subset M_0$ *be a closed set such that* f *is smooth in a neighborhood of* A. *Given* $\varepsilon: M \to R_+$ *there exists a prismatic* ε*-isotopy* ψ_t *of* M *such that*

(a) $\psi_t(\Delta) = \Delta$ *for each simplex* Δ *of* K, *and* $\psi_t|\Delta: \Delta \to \Delta$ *is a diffeotopy of* Δ;

(b) ψ_t *is fixed on* $M - U$ *and on a neighborhood of* A;

(c) $f \circ \psi_1 : \hat{K} \to R$ *is smooth of rank* 1 *in a neighborhood of* M_0;

(d) $f \circ \psi_1 : \hat{K} \to R$ *is* X-*regular.*

Proof. Let $\delta : \hat{K} \to R_+$. By 5.8 there exists a map $g : \hat{K} \to R$ such that

g is X-regular;

g simplicially smooth;

g is smooth in a neighborhood of M_0;

$g = f$ on $\hat{K} - U$ and in a neighborhood of A;

$|g - f| < \delta$; g and f are X-coregular.

Let $\psi_t : \hat{K} \to \hat{K}$ be the X-homotopy relating f to g. Extend ψ_t to be the identity outside $|K|$. If δ is small enough ψ_t will be an ε-isotopy, prismatic by 6.4b.

The following variation of 6.11 will be used in the proof of 8.1. To state it we define the *link* L(v) of a vertex v of a complex C to be the subcomplex:

$$L(v) = \{\sigma \epsilon St(v) \mid v \notin \sigma\} .$$

Thus $|St(v)|$ is the join of v and $|L(v)|$.

6.12 LEMMA. *Let* v *be a vertex of a complex* C *and* $g, f : St(v) \to R$ *simplicially affine maps. Suppose* $g(v) = f(v) = 0$ *and* $g(x) > 0$, $f(x) > 0$, *if* $x \epsilon L(v)$. *(Hence* $g^{-1}(0) = f^{-1}(0) = v$.) *Given a neighborhood* V *of* v *in* $|St(v)|$, *there exists an isotopy* ψ_t *of* $|St(v)|$ *such that* $\psi_0 = 1$ *and:*

(a) ψ_t *is fixed on* $|St(v)| - v$ *and also on* $\{x \epsilon |S(v)| : f(x) = G(x)\}$;

(b) ψ_t *maps each simplex of* St(v) *into itself by a diffeotopy;*

(c) $g\psi_1 = f$ *in a neighborhood of* v.

Proof. Let $<0,1> \subset R$ be the open interval $\{x \epsilon R \mid 0 < x < 1\}$. Let $\lambda(s,t) : I \to I$ be a family of diffeomorphisms indexed by $(s,t) \epsilon <0,1> \times <0,1>$ such that:

$\lambda(s,t)$ maps $[0,s]$ linearly onto $[0,t]$;

$\lambda(t,t) = 1_I$;

$\lambda(s,t) : x \mapsto x$ if $\max\{s,t\} \leq 2x - 1$;

the map $(s,t,x) \mapsto \lambda(s,t)(x)$ is smooth in (s,t,x).

Choose $\varepsilon \in \langle 0, \frac{1}{2} \rangle$ so that if $x \in L(v)$ then

$$[0, 2\varepsilon] \subset f(V \cap [v,x]) \cap g(V \cap [v,x]) .$$

Let $\gamma_x : [0,1] \to [v,x]$ be the affine map such that $0 \mapsto v$ and $1 \mapsto x$. Define $a_x, b_x \in \langle 0, 1 \rangle$ by

$$f\gamma_x(\varepsilon) = a_x, g\gamma_x(\varepsilon) = b_x .$$

Define ψ_t by

$$\psi_t \mid [v,x] = \gamma_x \circ \lambda(a_x, (1-t)a_x + tb_x) .$$

Then ψ_t has the required properties.

§7. *Smoothing submanifolds*

We digress from the proof of the main theorem in order to prove some applications of the results of the last section. They will be used in Section 10 for the proof of the main theorem and again in Section 12.

If M' is a PL manifold and $M \subset M' - \partial M'$ a PL submanifold of codimension k we call M *flat* if M has a PL collar in M'; that is, M has a neighborhood $E \subset M'$ such that $(E, M) \equiv (M \times R^k, M \times 0)$. If M has an open cover by sets $U_i \subset M$ which are flat submanifold of M', then M is *flat* in M.

Theorem 7.4 below says, in a strongly relative form, that a locally flat submanifold M of codimension 1 in a smoothed PL submanifold M'_α can be pushed onto a smooth submanifold by arbitrarily small PD isotopies of M'_α. In 7.5 a similar result is proved for flat submanifold of arbitrary codimension. Because the theorems are relative in form, it follows that the induced smoothing of M is unique up to concordance and therefore, after the main theorem is proved, it is unique up to isotopy.

First, to explain the main ideas, the absolute theorem is proved.

7.1 THEOREM. *Let* M' *be a* PL *manifold and* $M \subset M' - \partial M'$ *a flat* PL *submanifold. If* α *is a smoothing of* M', *there is a* PD *isotopy* $h_t : M' \to M'_\alpha$ *such that* $h_1(M) \subset M'_\alpha$ *is a smooth submanifold. Moreover, if* M *is compact, given any neighborhood* $N \subset M'$ *of* M *and any* $\varepsilon > 0$, h_t *can be chosen to be an* ε*-isotopy fixed on* $M' - N$.

Proof. By induction on $k = \dim M' - \dim M$. The last statement of the theorem will be left to the reader, since 12.1 below includes all of 7.1 as a special case.

We may replace (M', M) by $(M \times \mathbf{R}^k, M \times 0)$ since M is flat in M'. Suppose $k = 1$. Give $(M \times \mathbf{R})_\alpha$ a smooth triangulation C so that the projection $\pi_2 : M \times \mathbf{R} \to \mathbf{R}$ is simplicially affine. Let $y \in \mathbf{R}$ be such that $M \times y$ contains no vertices. There is a PL isotopy $f_t : M - \mathbf{R} \to M \times \mathbf{R}$ taking $M \times 0$ onto $M \times y$, for example, $f_t(x, s) = (x, s + ty)$. By 6.8 (applied to the complex $K = \{\sigma \in C | \sigma \cap M \times y \neq \emptyset\}$) $\pi_2 : (M \times \mathbf{R})_\alpha \to \mathbf{R}$ is regular on a neighborhood of $M \times y$. By 6.5 there is a PD isotopy $g_t : M \times \mathbf{R} \to (M \times \mathbf{R})_\alpha$ such that $g_1(M \times y)$ is a smooth submanifold. The PD isotopy $h_t = g_t f_t$ takes $M \times 0$ onto a smooth submanifold.

If $k > 1$, write $M \times \mathbf{R}^k = (M \times \mathbf{R}^{k-1}) \times \mathbf{R}$. Since the theorem is proved for $k = 1$, we may assume $M \times \mathbf{R}^{k-1}$ is a smooth submanifold of $[(M \times \mathbf{R}^{k-1}) \times \mathbf{R}]_\alpha$. Moreover by uniqueness of collaring we can assume that α is a product smoothing:

$$[(M \times \mathbf{R}^{k-1}) \times \mathbf{R}]_\alpha = (M \times \mathbf{R}^{k-1})_\beta \times \mathbf{R} .$$

By the induction hypothesis there is a PD isotopy $h_t : M \times \mathbf{R}^{k-1} \to (M \times \mathbf{R}^{k-1})_\alpha$ such that $h_1(M \times 0)$ is a smooth submanifold. Then

$$h_t \times 1_{\mathbf{R}} : (M \times \mathbf{R}^{k-1}) \times \mathbf{R} \to (M \times \mathbf{R}^{k-1})_\alpha \times \mathbf{R}$$

is the required PD isotopy. This completes the proof of 7.1.

The PD isotopy h_t constructed in the proof is not generally prismatic, since the PL isotopy f_t taking $M \times 0$ onto $M \times y$ is not generally prismatic.

7.2 THEOREM. *Let* M *and* N *be* PL *manifolds, with* $\partial N = \emptyset$. *If* $M \times N$ *has a smoothing, then both* M *and* N *have smoothings.*

Proof. First suppose $\partial M = \emptyset$. Let $U \subset N$ be PL homeomorphic to $\mathbf{R}^n$. Then $M \times \mathbf{R}^n$ has a smoothing; by 7.1, $M \times 0$ has a smoothing. Thus we find that M has a smoothing.

If $\partial M = \emptyset$, let $M_1 \subset M - \partial M$ be a PL homeomorphic to M. Then $M - \partial M$ has a smoothing α. Since ∂M_1 is collared in $M - \partial M$, we apply 7.1 to find a PD homeomorphism $f : M - \partial M \to (M - \partial M)_\alpha$ such that $f(\partial M_1)$ is a smooth submanifold. Therefore $f(M_1)$ is a smooth submanifold. Hence M is smoothable.

To see that N has a smoothing, observe that $N \times (M - \partial M)$ is smoothable; hence N is smoothable by what has been already proved. This completes the proof of 7.2.

We can now prove that the smoothability of a PL manifold depends only on its PL tangential homotopy type. It is known (see e.g., [6]) that smoothability is not an invariant of homotopy type.

7.3 THEOREM. *Let* M_0 *and* M_1 *be* PL *manifolds without boundaries, having the same proper* PL *tangential homotopy type. If* M_0 *has a smoothing, so has* M_1.

Proof. The hypothesis means there is a proper homotopy equivalence $f : M_0 \to M_1$ such that $f^*T(M_1) \approx T(M_0)$ where $T(M_i)$ is the PL tangent bundle. In [4] it is proved that this implies $M_0 \times \mathbf{R}^q \equiv M_1 \times \mathbf{R}^q$ for some $q \geq 0$. Therefore if M_0 is smoothable, so $M_1 \times \mathbf{R}^q$, and M_1 is smoothable by 7.2.

7.4 THEOREM. *Let* M' *be a* PL *manifold and* $M \subset M'$ *a locally flat* PL *submanifold of codimension* 1. *Let* α *be a smoothing of* M'. *Let* $N \subset M'$ *and* $K \subset M \cap N$ *be closed sets such that* $K \cap \mathrm{bd}\, N$ *has a neighborhood in* M *which is a smooth submanifold of* M'_α. *For every* $\varepsilon : M' \to R_+$ *there is a* PD ε*-isotopy* $f_t : M' \to M'_\alpha$ *such that*

(a) K *has a neighborhood* $W \subset M$ *such that* $f_1(W) \subset M'_\alpha$ *is a smooth submanifold*

(b) $f_t | M' - N = 1$.

In other words α *is* ε*-isotopic* rel $M' - N$ *to a smoothing* β *of* M' *having* W *as a smooth submanifold.*

Proof. We shall use the following lemma; it says, in a strongly relative form, that the graph of a PL map can be pushed onto the graph of a smooth map.

7.5 LEMMA. *Let* U *be a* PL *manifold and* α *a smoothing of* U. *Let* $N_1 \subset U$ *and* $L \subset N_1$ *be compact sets. Let* $0 < b < a$. *Suppose* $\lambda : U \to [-b, b]$ *is a* PL *map which is constant on each component of* $L \cap \mathrm{bd}\, N_1$. *Given* $\varepsilon > 0$ *there exists: a* PD ε*-isotopy* $h_t : U \times R \to U_\alpha \times R$; *a neighborhood* $A \subset N_1$ *of* L; *and a smooth map* $\lambda_0 : A \to \langle -a, a \rangle$, *such that*

(a) h_1 (*graph* $\lambda | A$) = *graph* λ_0;

(b) h_t *is fixed on* $U \times R - N_1 \times [-a, a]$;

(c) h_t *preserves the projection* $\pi_1 : U \times R \to U$.

The proof of the lemma, based on a smooth approximation to λ, is left to the reader.

Now consider the special case of 7.4 where M is flat in M', N is compact, and $\partial M = \emptyset$ so that $M \subset M' - \partial M'$. Call this the *first special* case. We may replace (M', M) by $(M \times R, M \times 0)$ and N by $N_0 \times [-a, a]$ where $N_0 \subset M$ is a compact set containing K and $0 < a < \varepsilon/8$.

By uniqueness of collaring (13.10) we may assume that $\alpha|U \times R$ is a product smoothing.

Give $(M\times R)_\alpha$ a smooth triangulation such that $\pi_2 : M \times R \to R$ is simplicially affine. Let $0 \le y < b < \varepsilon/8$ be chosen so that $M \times y$ contains no vertices. Observe that $U \times y \subset (M\times R)_\alpha$ is smooth, since $\alpha|U \times R$ is a product.

Let $\lambda : M \to [0, b]$ be a PL map such that

(1) $\lambda = 0$ on an open neighborhood of $M - \operatorname{int} N_0$

(2) $\lambda = y$ in an open neighborhood B of $K - U$.

There is a PL isotopy of $M \times R$, fixed outside of $N_0 \times [a, a]$ which carries $M \times 0$ onto the graph of λ. Moreover this isotopy can be chosen to preserve projection on M, and hence it is an $\varepsilon/4$-isotopy. By Lemma 7.2 there is a PD $\varepsilon/4$-isotopy of $(M\times R)_\alpha$ fixed on $M \times R - U \times [-a, a]$, carrying the graph of λ onto the graph $\lambda_0 : M \to [0, a]$ where

$\lambda_0 = 0$ in an open neighborhood of $M - \operatorname{int} N_0$;

λ_0 is smooth in an open neighborhood $U_1 \subset U$ of $K \cap \operatorname{bd} N_0$;

$\lambda_0 = y$ in a compact neighborhood $E \subset N_0$ of $K - U_1$.

Observe that λ_0 is smooth in a neighborhood of K. (This follows from 7.2 by letting $N_1 \subset U \cap N_0$ be a compact set containing $(K \cap U) - B$, where B is from (2).) Combining these two isotopies gives a PD $\varepsilon/2$-isotopy $\phi_t : M \times R \to (M\times R)_\alpha$ such that ϕ_t is fixed outside $N \times [a, a]$. Then

$$\pi_2 : (M\times R)_\alpha \to R$$

is regular on $\phi_1(N_0 \times 0)$ and smooth in a neighborhood of $\phi_1(\operatorname{bd}(N_0 \times [-a, a]) \cap K)$, by 6.9, since $M \times y$ contains no vertices. Put $\beta = \phi_1 * \alpha$. Then $\pi_2 : (M\times R)_\beta \to R$ is regular on $N_0 \times 0$ and smooth in a neighborhood of $(K \cap \operatorname{bd} N_0) \times 0$. By 6.10 there is a PD $\varepsilon/2$-isotopy $\psi_t : M \times R \to (M\times R)_\beta$ fixed outside of $N_0 \times [-a, a]$ such that

(3) $\pi_2\psi_1 : (M\times R)_\beta \to R$ is smooth of rank 1 in a neighborhood of $N_0 \times 0$.

Define a PD ε-isotopy $f_t : M \times \mathbf{R} \to (M \times \mathbf{R})_\alpha$ by

$$f_t = \begin{cases} \phi_{2t} & \text{if } 0 \leq t \leq \frac{1}{2} \\ \phi_1 \psi_{2t-1} & \text{if } \frac{1}{2} \leq t \leq 1 . \end{cases}$$

Then f_t is fixed outside $N_0 \times [-a, a]$ and N_0 has a neighborhood $W \subset M$ such that $f_1(W \times 0) \subset (M \times \mathbf{R})_\alpha$ is smooth. This is because $f_1(N_0 \times 0) = \phi_1 \psi_1(N_0 \times 0)$; by (3), N_0 has a neighborhood $W \subset M$ such that $\psi_1(W \times 0)$ is smooth in β; and $\phi_1 : (M \times \mathbf{R})_\beta \to (M \times \mathbf{R})_\alpha$ is a diffeomorphism. This finishes the proof of the first special case of 7.4.

Now consider the *second special case*: N compact and M' flat in M. Then by hypothesis $\partial M = M \cap \partial M'$. There exists a collaring of the pair $(\partial M', \partial M) \subset (M', M)$ (see [24]; in our application this condition is obviously satisfied). Thus $\partial M'$ has a neighborhood in M' parametrized by $(\partial M') \times [0, \infty>$. By what we have already proved we may assume that $K \cap \partial M$ has a neighborhood $V \subset \partial M$ which is smooth in $(\partial M')_{\partial \alpha}$. Let $E \subset \partial M'$ be an open neighborhood of $(K - U) \cap \partial M'$.

By uniqueness of collaring, we may perform a small PD isotopy of M'_α fixed on $\partial M'$ and outside N, which takes $\alpha | E \times [0, b]$ into a product smoothing, for some $b > 0$.

We have reduced the second special case to *the third special case*: M is flat in M', N is compact, and $K \cap \partial M'$ has a neighborhood $L \subset M'$ such that $L \cap M$ smooth in M'_α. (For example $L = E \times [0, b>$ in the notation above.)

Let $L_1 \subset L_0 \subset L$ be open sets such that $\mathrm{cl}(L_1) \subset L_0$, $\mathrm{cl}(L_0 \subset L$. Put

$$M'_* = M' - L_1$$

$$M_* = M - L$$

$$N_* = N - L_0$$

$$K_* = K - L_0 .$$

Then $M_* \subset M'_* - \partial M'_*$, N_* is compact, and $K_* \cap \operatorname{bd} N_*$ has a neighborhood in M_* which is smooth in $(M'_*)_\alpha$. Also M_* is flat in M'_*. We apply the first special case to (M'_*, M_*, N_*, K_*) and find a small PD isotopy g_{*t} of $(M'_*)_\alpha$, fixed on $M'_* - N_*$, taking a neighborhood $W_* \subset M_*$ of K_* onto a smooth manifold of $(M'_*)_\alpha$. Extend g_{*t} to a PD isotopy g_t of M'_α fixed outside N_*. Then g_t is fixed on $M' - N \subset M'_* - N_*$. Put

$$W = W_* \cup (L \cap M) \subset M .$$

This is a neighborhood of K in M, and $g_1(W)$ is the union of two open subsets, $g_{*1}(W_*) \cup (L_0 \cap M)$ each of which is a smooth submanifold of M'_α. This proves the third and the second special cases of 7.4.

The general case is proved by triangulating M' so finely that each simplex σ lies in a neighborhood of M which is flat in M'. Assuming inductively that the (i–1)-skeleton K_i of M has a neighborhood $A_i \subset M$ which is a smooth submanifold of M'_α for each i-simplex σ, let $M_\sigma \subset M$ be a flat neighborhood of $K_\sigma = \sigma - A_i$ so that $M_\sigma \cap M_\tau = \emptyset$ if σ and τ are disjoint i-simplices. Then apply the special case simultaneously to all the neighborhood M_σ. The details are left to the reader. (Compare Section 11 for a similar globalization.)

7.6 THEOREM. *Let M' be a PL manifold with a smoothing α. Let $M \subset M' - \partial M'$ be a closed PL submanifold; let $N \subset M'$ and $K \subset M \cap N$ be closed sets. Suppose:*

(a) *K has a neighborhood $N_0 \subset M$ such that N_0 is flat in M' and*

(b) *$K \cap \operatorname{bd} N$ has a neighborhood $U \subset M$ such that $U \subset M'_\alpha$ is a smooth submanifold. Then for any $\varepsilon : M' \to R_+$ there is a PD ε-isotopy $f_t : M' \to M'_\alpha$ such that:*

(c) *K has a neighborhood $W \subset M$ such that $f_1(W) \subset M'_\alpha$ is smooth;*

(d) *f_t is fixed on $M' - N$.*

Proof. By induction on $k = \dim M' - \dim M$, using 7.4 to start the induction if $k = 1$. The details are left to the reader.

Remark:

Notice that 7.6 is not a generalization of 7.4. In the latter the submanifold was assumed only locally flat, whereas in 7.6 we assumed the submanifold was flat where we wanted to smooth it. In codimension 1 we have uniqueness of collaring at our disposal. We could not expect a straightforward generalization of 7.1 to higher codimensions, since it would prove that every PL manifold could be smoothed. (For counterexamples see Smale [19], Kervaire [8].) The essence of uniqueness of collaring is that a submanifold of codimension one has a PL normal bundle which can be given a compatible vector bundle structure. The proper generalization of 7.4 is to PL submanifolds having such normal bundles: they can be isotoped onto smooth submanifolds. (See Lashof-Rothenberg [10]. Rourke-Sanderson [18].) The proof of this follows from 7.3 as soon as the proper definitions are made. (See Part II.)

We are now in a position to prove a weak form of the $M \times \mathbf{R}$ theorem referred to in the introduction:

7.7 THEOREM. *Let* α *be a smoothing of* $M \times \mathbf{R}$. *Then there exists a smoothing* β *of* M *such that* $(M\times\mathbf{R})_\alpha$ *is concordant to* $M_\beta \times \mathbf{R}$. *Such a smoothing* β *is unique up to concordance.*

Proof. By 7.4 we may assume that $M \times 0$ is a smooth submanifold, say $M_\beta \times 0 \subset (M\times\mathbf{R})_\alpha$. The normal bundle of $M_\beta \times 0 = M_\beta$ in $(M\times\mathbf{R})_\alpha$ is trivial; hence M_β has a smooth product neighborhood.

Let $f: M_\beta \times \mathbf{R} \to (M\times\mathbf{R})_\alpha$ be a smooth embedding which is the identity on $M \times 0$. By uniqueness of PD collars (13.10) there is a PD isotopy of f to the identity; let $F_t: M_\beta \times \mathbf{R} \to (M\times\mathbf{R})_\alpha$ be such an isotopy. Then $F_0^*\alpha = \beta \times \rho$ and $F_1^*\alpha = \alpha$. Therefore $G^*(\alpha \times \iota)$ is a concordance from $\beta \times \rho$ to α where

$$G : M \times \mathbf{R} \times I \to (M \times \mathbf{R}) \times I$$

is given by

$$G(x, y, t) = (F_t(x, y), t) .$$

This proves the existence of β.

To prove β unique, suppose $\beta_0 \times \rho$ and $\beta_1 \times \rho$ are concordant. Let γ be a smoothing of $M \times \mathbf{R} \times I$ which is a concordance from $\beta_0 \times \rho$ to $\beta_1 \times \rho$. Extend γ to a smoothing of $M \times \mathbf{R} \times \mathbf{R}$ so that $M_{\beta_1} \times \mathbf{R} \times [1, \infty\rangle$ and $M_{\beta_0} \times \mathbf{R} \times \langle -\infty, 0]$ are smooth submanifolds of $(M \times \mathbf{R} \times \mathbf{R})_\gamma$. Observe that the submanifold $N = M \times 0 \times \mathbf{R}$ is partly smooth, namely on

$$M \times 0 \times (\langle -\infty, 0] \cup [1, \infty\rangle) .$$

By 7.4 this partial smoothing can be extended to a smoothing η of N. If we restrict η to $M \times 0 \times [-2, 2]$ we obtain a concordance between $M_{\beta_0} \times 0 \times (-2)$ and $M_{\beta_1} \times 0 \times 2$. Therefore β_0 and β_1 are concordant.

7.8 COROLLARY. *The natural map from concordance classes of smoothings of* M *to concordance classes of smoothings of* $M \times \mathbf{R}^n$ *is bijective.*

§8. *Isolating singular points*

In this section we prove a result which shows that a smoothing of $M \times I$ is isotopic to a smoothing for which the projection $\pi_2 : M \times I \to I$ has an isolated singular set.

PL *nonsingular maps*

Let P be a PL manifold. A PL map $\lambda : P \to \mathbf{R}$ is called PL *nonsingular* if each point of $P - \partial P$ belongs to a coordinate chart $\phi : U \to \mathbf{R}^{m-1} \times \mathbf{R}$ such that $\lambda | U$ is the composition

$$U \xrightarrow{\phi} \mathbf{R}^{m-1} \times \mathbf{R} \xrightarrow{\pi_2} \mathbf{R} .$$

We call the pair (ϕ, U) a *local representation* of λ.

8.1 LEMMA. *Let* P *be* PL *manifold and* $\lambda : P \to \mathbf{R}$ *a* PL *nonsingular map which is constant on each component of* ∂P. *Let* C *be a triangula-*

tion of P *such that* f *is simplicially affine in* C. *Given* $\varepsilon : P \to R_+$, *there exists a positive function* δ *on the vertices of* C, *such that if* $\lambda' : C \to R_+$ *is a simplicially affine map such that* $|\lambda'(v) - \lambda(v)| < \delta(v)$ *for each vertex* $v \in C$ *and* $\lambda' = \lambda$ *on* ∂P, *then there is a* PL ε*-isotopy* $h_t : P \to P$ *such that* $\lambda = \lambda' h_1$. *Moreover* h_t *is fixed on* ∂P, *and on every simplex of* C *on which* $\lambda = \lambda'$. *In addition* λ' *is* PL *nonsingular; and for each vertex* v *of* C, $\mathrm{Lip}(\lambda' - \lambda) < \varepsilon(v)$ *on* $|\mathrm{St}(v)|$.

Proof. Let $v \in \partial P - P$ be a vertex of C. There is a local representation $\theta : |\mathrm{St}(v)| \to R^{m-1} \times R$ of λ. To see this, let $U \subset |\mathrm{St}(v)|$ be a neighborhood of v and $\phi : U \to R^{m-1} \times R$ a local representation of λ, so that

(1) $\pi_2 \phi = \lambda$

Choose $0 < k < 1$ so that if $x \in \mathrm{St}(v)$ then $kx + (1-k)v \in U$. Define

$$\phi_0 : |\mathrm{St}(v)| \to R^{m-1} \times R$$

by

$$\phi_0(x) = \phi(x) - \phi(v) ,$$

so that

(2) $\pi_2 \phi_0(x) = \lambda(x) - \lambda(v)$

and define

$$\theta : |\mathrm{St}(v)| \to R^{m-1} \times R$$

by

$$\theta(x) = k^{-1} \phi_0(kx + (1-k)v) + \phi(v) .$$

Then

$$\pi_2 \theta(x) = k^{-1} [\pi_2 \phi_0(kx + (1-k)v] + \pi_2 \phi(v) ,$$

since $\pi_2 : R^{m-1} \times R \to R$ is linear,

$$= k^{-1} [\lambda(kx + (1-k)v) - \lambda(v)] + \lambda(v)$$

by (1) and (2),

$$= k^{-1}[k\lambda(x)+(1-k)\lambda(v)-\lambda(v)] + \lambda(v)$$

because λ is affine on each simplex,

$$= k^{-1}[k\lambda(x)-k\lambda(v)] + \lambda(v)$$

$$= \lambda(x) .$$

Suppose then that $\theta : |St(v)| \to R^{m-1} \times R$ is a PL embedding such that $\pi_2\theta = \lambda$. Let C' be a subdivision of C in which θ is simplicially affine. Given $s > 0$, let $r > 0$ be so small that if $w \in |St_C(v)| - |\partial St_C(v)|$ is a vertex of C' and $|y-\lambda(w)| \leq r$, there exists a point

$$z \in \text{int}\, \theta(|St_C(v)|) \cap (R^{n-1} \times y)$$

such that $|z-\theta(w)| < s$.

Let $\lambda' : St_C(v) \to R$ be a simplicially affine map such that $\lambda' = \lambda$ on $\partial St_C(v)$. Thus λ' and λ agree on every vertex of $St_C(v)$ except v.

Suppose $|\lambda'-\lambda| < r$. Let

$$\theta' : |St_C(v)| \to R^{m-1} \times R$$

be a PL map which is simplicially affine in the subdivision L of $St_C(v)$ induced by C' such that for each vertex $w \in L$ we have

(3) $\theta'(w) \in \text{int}\, \theta|St_C(v)| \cap R^{m-1} \times \lambda'(y)$, and

(4) $|\theta'(w) - \theta(w)| < s$.

Observe that

$$\pi_2\theta' = \lambda' ,$$

since this is true for each vertex of L by (3), and θ' and λ' are simplicially affine in L, while π_2 is a linear map. It is easy to see that as $s \to 0$, (4) implies that $|\theta'-\theta| \to 0$; also $|\text{Lip}(\theta'-\theta)| \to 0$ on $|St_C(v)|$ and therefore θ' will be an embedding if s is sufficiently small. Moreover given $q > c$ we can choose s so small that the PD homotopy, fixed on $\partial St_C(v)$,

$$(1-t)\theta + t\theta' : |\mathrm{St}_C(v)| \to R^{m-1} \times R$$

will be a PD isotopy, and can be approximated by a PL q-isotopy

$$\theta_t : |\mathrm{St}_C(v)| \to R^{m-1} \times R$$

such that $\theta_0 = \theta, \theta_1 = \theta'$ and θ_t is fixed on $|\partial\,\mathrm{St}_C(v)|$. Define a PL isotopy h_t of $|\mathrm{St}_C(v)|$ by $h_t = \theta_t^{-1}\theta_0$. Extend h_t to be fixed on $P - |\mathrm{St}_C(v)|$. Then $h_t = 1$ on $|\partial\,\mathrm{St}_C(v)|$, and $\lambda' h_1 = \lambda'\theta_1^{-1}\theta_0 = \pi_2\theta_1\theta_1^{-1}\theta_0 = \lambda$. Observe that λ' is PL nonsingular since h_1 exists, and it is easy to see that $\mathrm{Lip}(\lambda'-\lambda) \to 0$ and $|\lambda'-\lambda| \to 0$ as $s \to 0$ since $\pi_2 : R^{2n-1} \times R \to R$ is linear.

We have proved 8.1 in the special case where $\lambda' = \lambda$ on every vertex of C except v. The general case follows by an obvious globalization passing from λ to λ' by changing λ successively over an enumeration of the vertices of C. The reader can supply the details.

Isolating singularities by separating vertices

8.2 THEOREM. *Let* M *be a smooth manifold,* P *a* PL *manifold and* $T : P \to M$ *a* PD *homeomorphism. Let* $\lambda : P \to R$ *be a* PL *nonsingular map which is constant on each component of* ∂P. *Let* $N \subset M$ *be a closed set such that* $\lambda T^{-1} : M \to R$ *is regular on* bd N. *Given* $\varepsilon : P \to R_+$ *there is a* PL *ε-isotopy* h_t *of* P *such that*

(a) *the map* $\lambda h_1 T^{-1} : M \to R$ *has only isolated singular points in a neighborhood of* N;

(b) $\lambda h_1 T^{-1}$ *is regular on* bd N

(c) h_t *is fixed on* $(P - T^{-1}N) \cup \partial P$.

Proof. Since the set of regular points of λT^{-1} is open, bd N has a neighborhood $U \subset M$ on which λT^{-1} is regular.

Give P a triangulation C in which λT^{-1} is simplicially affine and so fine that every simplex meeting $T^{-1}(\mathrm{bd}\,N)$ lies in U. Let V be the set of vertices of C lying in

$$P - (\partial P \cup |\mathrm{St}(T^{-1}\,\mathrm{bd}\,N|)\ .$$

For each vertex v of C choose $\lambda'(v) \in \mathbf{R}$ such that $\lambda'(v) \neq \lambda'(w)$ if $v \neq w$ and $v \in V$. This defines a simplicially affine map $\lambda' : C \to \mathbf{R}$. If the numbers $|\lambda'(v) - \lambda(v)|$ are sufficiently small, the theorem will be satisfied by 8.1 and 5.4.

With a view toward the main theorem, we apply the last result to smoothings of $M \times I$.

8.3 THEOREM. *Let* M *be a* PL *manifold without boundary and* α *a smoothing of* $M \times I$. *Let* $N \subset M$ *be a closed set such that* α *is a product in a neighborhood of* $(\mathrm{bd}\,N) \times I$. *Given* $\varepsilon : M \times I \to \mathbf{R}_+$, α *is* ε*-isotopic* rel $[(M-N)\times I \cup \partial M \times \{0,1\}]$ *to a smoothing* β *such that* $\pi_2 : (M\times I)_\beta \to I$ *has only isolated singular points in a neighborhood of* $N \times I$, *and* π_2 *is smooth of rank* 1 *on a neighborhood of* $N \times \{0,1\}$.

Proof. By uniqueness of collaring (13.10) we may assume that each point of $N \times 0$ has a neighborhood $U \times [0,r]$ with $r > 0$ such that $\alpha | U \times [0,r]$ is a product; similarly for $N \times 1$. Therefore we may assume π_2 smooth of rank 1, and hence regular, in a neighborhood of $N \times \{0,1\}$. The theorem is now a consequence of 8.2 replacing M of 8.2 by $M \times \langle 0,1 \rangle$, N of 8.2 by $N \times \langle 0,1 \rangle$ etc.

§9. *Removing isolated singularities*

In this section we prove 4.2, the simplified form of the main theorem. It is proved together with 9.1 by double induction, as follows:

$$4.2\,(n-1) \Longrightarrow 9.1\,(n+1) \Longrightarrow 4.2\,(n)$$

starting with 4.2 (–1) which is vacuous, or 9.1 (1) which is trivial.

Removable singularities

Throughout Section 9, K denotes a complex which is the star of the vertex 0 in a rectilinear triangulation of R^m. A *simplicially smooth embedding* $T: K \to R^m$, $T(0) = 0$ is given; this means that $T: |K| \to R^m$ is an embedding and $T|\sigma: \sigma \to R^m$ is smooth of rank m for each m simplex $\sigma \in K$. For example T may be part of a smooth triangulation.

We shall denote by $\pi: K \to R$ a simplicially affine map which is PL nonsingular (Section 8), and such that $\pi T^{-1}: T|K| \to R$ has an isolated singular point at $0 \in T|K|$. It follows that πT^{-1} is regular on $T|K| - 0$.

We say that (π, T, K) has a *removable singularity* if there exists a PD isotopy $T_t: |K| \to R^m$ of T such that

(1) $T_t = T_0$ in a neighborhood of $|\partial K|$ and

(2) $\pi T_1^{-1}: T|K| \to R$ is regular.

Note that πT_t^{-1} is not required to be regular for $0 < t < 1$ except implicitly by (1) on a neighborhood of $|\partial K|$. We call T_t a *regularizing isotopy* for (π, T, K).

9.1 THEOREM. *Let* $T: K \to R^m$ *and* $\pi: K \to R$ *be as above. Then* (π, T, K) *has a removable singularity.*

Some lemmas

We prove some lemmas whose purpose is to allow T in 9.1 to be replaced by isotopic embeddings and K to be replaced by "substars." The difficulty is that if $T_1: K \to R^m$ is prismatically isotopic to $T, \pi T_1^{-1}$ need not be regular on $T_1|K| - 0$. One convenient sufficient condition for this is that $\pi(v) \neq \pi(0)$ for boundary vertices of K. Another is that $T_1 = f_1 T$ where f_t is a *diffeotopy*; the inductive hypothesis of the main theorem is used in 9.12 and 9.13 to construct such a diffeotopy. Still another method of making πT_1^{-1} regular on $T_1|K| - 0$ is to make sure that T_1 is sufficiently Lipschitz close to T.

9.2 LEMMA. *Let* L *be the star of* 0 *in a triangulation of* $|K|$. *If* $(\pi|L, T|L, L)$ *has a removable singularity at* 0, *so has* (π, T, K).

Proof. Let $S_t : |L| \to R^m$ be a regularizing isotopy for $(\pi|L, T|L, L)$. Define $T_t : |K| \to R^m$ to equal S_t on $|L|$ and T on $|K| - |L|$. Then T_t is a regularizing isotopy for (π, T, K).

9.3 LEMMA. *Let* L *be the star of* 0 *in a subdivision of* K. *Let* $\pi' : L \to R$ *be a simplicially affine map such that*

(a) *there is a* PL *isotopy* h_t *of* $|K|$, *fixed on* $|K| - |L|$ *such that* $\pi = \pi' h_1$

(b) $(\pi', T|L, L)$ *has a removable singularity.*

Then (π, T, K) *has a removable singularity.*

Proof. Let $S_t : |L| \to R^m$ be a regularizing isotopy for $(\pi', T|L, L)$. Extend S_t to equal T on $|K| - |L|$. Put $T_t = S_t h_t : |K| \to R^m$. Then $\pi T_1^{-1} = \pi h_1^{-1} S_1^{-1} = \pi' S_1^{-1}$, which is regular. Therefore T_t is a regularizing isotopy for (π, T, K).

9.4 LEMMA. *Given* $0 < s_1 < s_2 < 1$, *there exists* $\varepsilon > 0$ *with the following property. Suppose there exists* $T_1 : |K| \to R^m$, *a* PD *embedding such that*

(a) T_1 *is* PD *isotopic to* T rel *a neighborhood of* $|\partial K|$

(b) πT_1^{-1} *is regular on* $\operatorname{int} |s_1 K| - 0$

(c) $\operatorname{Lip}(T_1 - T) < \varepsilon$ *and* $|T_1 - T| < \varepsilon$ *on* $|K| - |s_1 K|$.

Then T_1 *is regular on* $|K| - 0$.

Proof. It suffices to know that T_1 is regular on $|K| - |s_1 K|$. If ε is small enough this follows from 5.4 (setting $E = T|K|, \phi = \pi T_1^{-1}, h = T T_1^{-1}$) and the observation that $|T T_1^{-1} - 1| \to 0$ and $\operatorname{Lip}(T T_1^{-1} - 1) \to 0$ as $|T - T_1| \to 0$ and $\operatorname{Lip}(T - T_1) \to 0$.

9.5 LEMMA. *Suppose* $\pi(v) \neq \pi(0)$ *for every vertex* $v \in \partial K$. *Let* $T_t : K \to \mathbf{R}^m$ *be a prismatic isotopy of* T. *Then* πT_1^{-1} *is regular on* $\mathrm{int}\, T_1|K| - 0$.

Proof. Follows from 6.9.

Some special isotopies

Let $A \subset B \subset |K|$ be any subsets. A map $f : B \to \mathbf{R}^m$ is called *radial* on A if for each interval $[tx, x] \subset A$ with $t \in I$, $f[tx, x] = [sf(x), f(x)]$ for some $s \in I$. If in addition $s = t$, that is if $f(tx) = tf(x)$ whenever $t \in I$ and $[tx, x] \subset A$, we call f *affinely radial* on A.

9.6 LEMMA. *Suppose* $T|K| = D^m \subset \mathbf{R}^m$, *the unit ball. Assume* T *is affinely radial in an neighborhood* U *of* $|\partial K|$ *in* $|K|$. *Let* $0 < s < 1$ *be such that* $|\partial sK| \subset U$. *Let* $V \subset U$ *be a neighborhood of* $|\partial sK|$. *Let* $g_t : \partial sK \to \partial sD^m$ *be a prismatic isotopy of* $T|\partial sK = g_0$. *Then* g_t *extends to a prismatic isotopy* T_t *of* T *which is fixed outside* V *and radially affine in a neighborhood of* $|\partial sK|$. *If* g_t *preserves* $\sigma \cap |\partial sK|$ *for each* $\sigma \in K$ *and* $t \in T$, *then* T_t *preserves* σ.

Proof. Let $0 < \varepsilon < \min\{s, 1-s\}$ be such that $|\partial rK| \subset V$ if $|r - s| < \varepsilon$. Choose a smooth map $\lambda : [s-\varepsilon, s+\varepsilon] \to I$ equal to 1 on a neighborhood of s and 0 on a neighborhood of $\{s-\varepsilon, s+\varepsilon\}$. For each $x \in |\partial K|$ and $0 \leq t \leq 1$, $0 \leq r \leq 1$ define

$$T_t(rx) = \begin{cases} f(x) & \text{if } r \notin [s-\varepsilon, s+\varepsilon] \\ r g_{\lambda(r)t}(sx) & \text{if } r \in [s-\varepsilon, s+\varepsilon]\,. \end{cases}$$

Then $T_t : K \to \mathbf{R}^{n+1}$ is the required isotopy.

9.7 LEMMA. *There is a prismatic isotopy* $T_t : K \to \mathbf{R}^m$ *of* T *which is fixed in a neighborhood of* $|\partial K|$, *such that for some neighborhood* U *of* 0 *in* $|K|$, $T_1|U \cap \sigma$ *is affine for each simplex* $\sigma \in K$ *containing* 0.

Proof. Define a homotopy $S_t : |K| \to R^m$ by

$$S_t(x) = \begin{cases} t^{-1}\, T(tx) & \text{if } 0 < t \leq 1 \\ D(T|\sigma)_0(x) & \text{if } t = 0 \end{cases}$$

where $\sigma \in K$ is a simplex containing 0 and x.

Let $\lambda : R^m \to I$ be a smooth map such that $\lambda = 1$ in a neighborhood of 0 and $\lambda = 1$ outside a neighborhood $V \subset |K|$ of 0. Let $r \geq 1$, and define

$$T_t(x) = \lambda(rx)\, S_{1-t}(x) + [1 - \lambda(rx)]\, T(x) \, .$$

If r is sufficiently large then T_t is the desired prismatic isotopy.

9.8 LEMMA. *There is a prismatic isotopy* $T_t : K \to R^m$ *which is fixed in a neighborhood of* $|\partial K|$ *and on* 0, *such that for some* $0 < s < 1, T_1|sK|$ *is a ball in* R^m *with center* 0, *and* T_1 *is affinely radial in a neighborhood of* $|\partial sK|$.

Proof. By 9.6 we may assume $T : K \to R^m$ is simplicially affine. For some $0 < s < 1$ let $B \subset \text{int}\, T|sK|$ be a ball with center 0. We leave it to the reader to construct a radial prismatic isotopy T_t carrying $s|K|$ onto B, having the desired properties.

Reduction of 9.1 to special cases

9.9 LEMMA. *In proving Theorem* 9.1, *it suffices to consider the special case where* $\pi(v) \neq \pi(0)$ *if* $v \in \partial K$ *is a vertex.*

Proof. Follows from 8.1, 9.3, and 9.4.

If $\pi : K \to R$ is as in 9.1 and T_t is a prismatic isotopy, then πT_1^{-1} is regular on $|K| - 0$ by 9.5. If also T_t is fixed on a neighborhood $|\partial K|$ it suffices to prove 9.1 for T_1. This allows us to specialize T still more:

9.10 LEMMA. *It suffices to prove 9.1 for the following special case:*

(a) $\pi(v) \neq \pi(0)$ *for every vertex* $v \in \partial K$

(b) $T|K| = D^m$

(c) T *is affinely radial in a neighborhood of* $|\partial K|$.

Proof. 9.9, 9.8 and 9.2.

9.11 LEMMA. *It suffices to prove 9.1 in the following special case:*

(a) $\pi(0) = 0$ *and* $\pi(v) \neq 0$ *if* $v \in \partial K$ *is a vertex*

(b) $T|K| = D^m$

(c) T *is affinely radial in a neighborhood of* $|\partial K|$.

(d) *The map* $\pi T^{-1} : D^m \to R$ *is smooth of rank* 1 *in a neighborhood in* D^m *of* $\{x \in \partial D^m | \pi T^{-1}(x) = 0\}$.

Proof. We may assume T as in 9.10. Let $U \subset |K|$ be an open set containing $|\partial K|$ on which T is affinely radial. Let $0 < r < 1$ be such that $\partial(rK) \subset U$. By (a) and (b),

$$T|rK| = rD^m = \text{the ball of radius } r \text{ in } R^m .$$

The boundary of rD^m is rS^{m-1}. Consider rS^{m-1} as a smooth manifold, smoothly triangulated by $T|\partial(rK) : \partial(rK) \to rS^{m-1}$. Put $f = \pi T^{-1}$: $rS^{m-1} \to R$. By 9.10(a) rS^{m-1} has no vertices in $M_0 = f^{-1}(0)$. By Lemma 6.10 there is a prismatic isotopy $h_t : rS^{m-1} \to rS^{m-1}$ such that $fh_1 : rS^{m-1} \to R$ is smooth of rank 1 in a neighborhood of M_0. By 9.6, the prismatic isotopy

$$h_t T : \partial(rK) \to rS^{m-1}$$

extends to a prismatic isotopy

$$T_t : K \to R^m$$

which is fixed outside of U, and such that T_1 is affinely radial in a neighborhood of $|\partial rK|$. Observe that since $T_1 \pi^{-1}$ is smooth on a neighborhood of M_0 in rS^{m-1}, the affine radiality of T_1 makes $T_1\pi^{-1}$ smooth on a neighborhood of M_0 in $\mathbf{R}^m$. By 9.5 and 9.2 it suffices to prove 9.1 for $(\pi|rK, T_1|rK, rK)$. This proves Lemma 9.11.

Use of the induction hypothesis 4.2(n–1)

9.12 LEMMA. *Assume that Theorem* 4.2 *is true for all manifolds* M *of dimension* n–1. *Let* $T: K \to D^{n+1} \subset \mathbf{R}^{n+1}$ *be as in* 9.11 *with* $m = n+1$. *Then the smooth submanifold*

$$B^{n-1} = \{x \in S^{n-1} | \pi T^{-1}(x) \geq 0\} \subset S^n$$

is diffeomorphic to the unit ball $D^n \subset \mathbf{R}^n$.

Proof. Observe that B^{n-1} is smoothly triangulated by $T|\partial K| \cap \pi^{-1}(0)$. Using the standing hypothesis that π is PL nonsingular it is easy to see that $|\partial K| \cap \pi^{-1}(0)$ is PL homeomorphic to an n–1 simplex. Now apply 4.3 to prove Lemma 9.12.

In the rest of this section we assume that 4.2(n–1) *is true, and that* $m = n+1$ *in* 9.1.

Let $\pi_{n+1}: \mathbf{R}^{n+1} \to \mathbf{R}$ denote the projection $\pi_{n+1}(x_1, \ldots, x_{n+1}) = x_{n+1}$.

9.13 LEMMA. *It suffices to prove* 9.1(n+1) *in the following special case*:

(a) $T|K| = D^{n+1}$

(b) $\pi T^{-1} = \pi_{n+1}$ *in a neighborhood* $E \subset D^{n+1}$ *of the equator* S^n.

(c) T *is affinely radial in a neighborhood of* $|\partial K|$.

Proof. Let T be as in 9.11, with $m = n+1$. Let $0 < s < 1$ be such that T is affinely radial in a neighborhood U of $|K| - \text{int}\,|sK|$. Let $V \subset \text{int}\,|K|$ be a compact neighborhood of $|\partial(sK)|$ in U. Observe that

$$sB^n = \{x \in \partial(sD^{n+1}) | \pi T^{-1}(x) \geq 0\}$$

and sB^n is a smooth submanifold of $\partial(sD^{n+1}) = sS^n$ diffeomorphic to B^n.

There is a diffeotopy of sS^n carrying sB^n onto sE^n_+ where E^n_+ is the northern hemisphere of $S^n : E^n_+ = \{x \in S^n | x_{n+1} \geq 0\}$. This diffeotopy extends to a diffeotopy ϕ_t of D^{n+1} which is fixed outside $T(V)$ such that ϕ_t is affinely radial in a neighborhood of $\partial(sD^{n+1})$.

Let $T_t = \phi_t T : K \to D^{n+1}$. This is a prismatic isotopy of T fixed in a neighborhood of $|\partial K|$ such that $T_1 |\partial sK| = sD^n_+$. Since ϕ_t is a diffeotopy, πT_1^{-1} is regular on $D^{n+1} - 0$. Also T_1 is affinely radial in a neighborhood of $|\partial(sK)|$.

Observe that $T_1((\partial sK) \cap \pi^{-1}(0)) = \partial(sD^{n+1}) \cap \pi_{n+1}{}^{-1}(0)$. In other words $\pi T_1^{-1}(0) \cap sS^n = sS^{n-1} = \pi_{n+1}{}^{-1}(0) \cap sS^n$. Moreover πT_1^{-1} is smooth of rank 1 in a neighborhood of sS^{n-1} in sS^n. By uniqueness of smooth collaring, there is a diffeotopy h_t of sS^n such that

(1) $\pi_{n+1} h_1 = \pi T_s^{-1}$

on a neighborhood of sS^{n-1} in sS^n.

We extend h_t to a diffeotopy of D^{n+1} in a way similar to that of the extension of ϕ_t above, and compose h_t with T_1 to find a prismatic isotopy $T_{1+t} : K \to D^{n+1}$ of T_1 with the following properties ($0 \leq t \leq 1$):

(2) $T_{1+t} = T_1$ in a neighborhood of $|\partial K|$

(3) $T_2 |sK| = sD^{n+1}$

(4) T_2 is affinely radial in a neighborhood of $|\partial(sK)|$

(5) $\pi T_2^{-1} : D^{n+1} \to R$ is regular on $D^{n+1} - 0$

(6) $\pi T_2^{-1} = \pi_{n+1}$ on a neighborhood of sS^{n-1} in sS^n.

This last follows from (1) since

$$\pi T_2^{-1} = \pi (h_1 T_1)^{-1} = \pi T_1^{-1} h_1^{-1} = \pi_{n+1} .$$

From (6) and (4) we get (7) $\pi T_2^{-1} = \pi_{n+1}$ on a neighborhood of sS^{n-1} in D^{n+1}.

By 9.3 it suffices to prove 9.1 for $(\pi | sK, T_2 | sK, sK)$. This proves 9.13.

PL *approximation and the Alexander trick*

We now complete the proof of 9.1 (n+1).

Let $T : K \to D^{n+1}$ be as in 9.13. Explicitly, we assume $\pi(0) = 0$ and:

(1) There exists $0 < q < 1$ such that T is affinely radial in a neighborhood of $|K| - \text{int}\,|qK|$.

(2) There exists $b > 0$ such that $\pi = \pi_{n+1}T$ on a neighborhood of $\pi^{-1}[-b, b] \cap (|K| - \text{int}\,|qK| = P$.

Let $q < r < s < 1$.

Given $\delta > 0$, there exists by a PD homeomorphism $T_1 : |K| \to D^{n+1}$ such that:

(3) $T_1 = T$ on $|K| - \text{int}\,|sK|$

(4) $T_1 : |rK| \to R^{m+1}$ is PL

(5) $|T_1 - T| < \delta$ and Lip $|T_1 - T| < \delta$

(6) T_1 is PD isotopic to T rel $|K| - \text{int}\,|sK|$

(7) $\pi_{n+1}T_1 = \pi_{n+1} = \pi$ on P.

If δ is sufficiently small, then from (1) we can easily obtain

(8) $T_1|rK|$ is strictly convex, and $0 \in \text{int}\, T_1|rK|$.

We now define a new PD homeomorphism

$$T_2 : |K| \to D^{n+1}$$

completely determined by T_1 and r, by declaring

(9) $T_2 = T$ on $|K| - \text{int}\,|rK|$

and

(10) $T_2 : |rK| \to R^{n+1}$ is affinely radial.

In other words, $T_2 \mid |rK|$ is the cone from 0 on $T_2 : |(\partial(rK)| \to R^{n+1}$. From (4) and (8) we have

(11) $T_2 \mid |rK|$ is a PL embedding.

Put

$$P \cap |\partial K| = Q = \{x \in |\partial K| : \pi(x) \leq b\} .$$

Let $C(Q) = \{tx \in |K| : t \in I \text{ and } x \in Q\}$ be the cone on Q.

It follows from (1) and (10) that

(12) $\pi = \pi_{n+1} T_2$ on $C(Q)$.

The "Alexander trick" (13.11) is the proof by "shrinking" that two PL homeomorphisms of a ball which agree on the boundary are PL isotopic rel the boundary. Applying this fact to T_2 and T_1 on $|rK|$, and using (6) we get,

If δ is sufficiently small then by 9.4

(14) πT_2^{-1} is regular on $T_2|K| - 0$.

By 9.3 it suffices to prove that $(\pi| \, |rK|, T| \, |rK|, |rK|)$ has a removable singularity. Since $T| \, |rK|$ enjoys properties (10), (11) and (12), 9.1(n+1) will follow from:

9.14 LEMMA. *Let K be the star of the vertex 0 of a rectilinear triangulation of* R^m. *Let* $\pi : K \to R$ *be a simplicially affine,* PL *nonsingular map with* $\pi(0) = 0$. *Let* $T : K \to R^m$ *be a simplicially affine embedding. Suppose* $\pi^{-1}(0) \cap |\partial K|$ *has a neighborhood* $Q \subset |\partial K|$ *such that* $\pi = \pi_m T$ *on* $C(Q)$. *Then* (π, T, K) *has a removable singularity.*

Proof. First observe that

(1) $\pi T^{-1} | T(\text{int } C(Q))$ is regular,

since $\pi = \pi_m T$ on $C(Q)$. Moreover

(2) πT^{-1} is regular on $T|K| - \pi_m^{-1}(0)$,

by 6.9. From (1) and (2) we see that

(3) πT^{-1} is regular on $T|K| - 0$.

We may assume Q is a subcomplex of ∂K, by replacing K by the cone on a subdivision of ∂K. Let $L = \{\sigma \epsilon \partial K | \sigma \not\subset Q\}$. Thus L = closure of $|\partial K| - |Q|$. Let $C(L) \subset K$ be the cone on L. We shall show there is a prismatic isotopy $\phi_t : C(L) \to \mathbf{R}^{m+1}$ of $T|C(L)$ such that

(4) $\pi \phi_1^{-1} = \pi_{m+1}$ in a neighborhood of 0 in $\phi_1|C(L))$

(5) ϕ_t is fixed on $C(L \cap Q)$ and in a neighborhood of $|L|$ in $|C(L)|$.

This will suffice to prove 9.14 for we then define $T_t : K \to \mathbf{R}^m$ to equal ϕ_t in $C(L)$ and T elsewhere. Then $\pi T_1^{-1} = \pi_{m+1}$ in a neighborhood of 0, and hence is regular there while πT_1^{-1} is regular on $T(|K| - \pi^{-1}(0))$ by 6.9.

We construct ϕ_t separately on $C(L_+)$ and $C(L_-)$, where

$$L_+ = \{\sigma \epsilon L | \pi(\sigma) \subset \mathbf{R}_+\}$$

$$L_- = \{\sigma \epsilon L | \pi(\sigma) \subset \mathbf{R}_-\} .$$

The existence of $\phi_t | L_+$ follows from 6.11, putting $C = C(L_+)$, $v = 0$, $L(v) = L_+$, $f = \pi$, $g = \pi_{m+1}$. Similarly for L_-. This completes the proof of 9.14.

We have proved the implication:

$$\text{Theorem } 4.2\,(n-1) \implies \text{Theorem } 9.1\,(n+1)$$

Proof of simplified form of main theorem

We now prove $9.1\,(n+1) \implies 4.2\,(n)$, which will complete the proofs of 9.1 and 4.2.

In 4.2 (n) we have a PL n-manifold M without boundary, a closed set $N \subset M$ of diameter d and a smoothing α of $M \times I$ which is a product in a neighborhood of $(\operatorname{bd} N) \times I$. Given $\varepsilon > 0$ we are required to find $\beta \epsilon S(M \times I)$ such that $\alpha \cong_{d+\varepsilon} \beta$ rel $M \times 0 \cup (M - N) \times I$ and which is a product in a neighborhood of $N \times I$. By 8.3 and an $\varepsilon/3$-isotopy we may assume that $\pi_2 : (M\times I)_\beta \to I$ is smooth of rank 1 in a neighborhood of $(\mathrm{bd}\, N\times I) \cup (N\times \mathrm{bd}\, I)$ and has only isolated singular points in a neighborhood W of $N \times I$.

By applying 9.1 (n+1) to a small neighborhood of each singular points of $\pi_2 : (W - \partial W)_\alpha \to I$, noting that $\dim W = n+1$, we can assume after an $\varepsilon/3$-isotopy that $\pi_2 : (M\times I)_\alpha \to I$ is regular in a neighborhood of $N \times I$.

By 6.6 there is an $\varepsilon/3 + d$ isotopy rel $M \times 0 \cup (M-N) \times I$ which fulfills the requirements of 4.2 (n).

§10. *Strongly relative form of the simplified main theorem*

10.1 THEOREM. *Let* M *be a* PL *manifold without boundary,* $N \subset M$ *a compact set, and* $K \subset N$ *a closed set. Let* α *be a smoothing of* $M \times I$ *which is a product in a neighborhood* $U \times I$ *of* $(K \cap \mathrm{bd}\, N) \times I$. *Given* $\varepsilon > 0$ *there exists a* PD *isotopy* $H_t : M \times I \to (M\times I)_\alpha$ *such that:*

(a) $H_t = 1$ *on* $M \times 0 \cup (M-N) \times I$;

(b) $H_1^*\alpha$ *is a product smoothing on a neighborhood of* K;

(c) $|\pi_2 H_t - \pi_2| < \varepsilon$ *where* $\pi_2 : M \times I \to I$ *is the projection.*

To explain the idea of the proof, suppose for simplicity that $K = N$. We may assume N is a PL submanifold. If $(\partial N) \times I \subset (M\times I)_\alpha$ happened to be a smooth manifold with a product smoothing, we could make α a product in a neighborhood of ∂N and apply 4.2. If $(\partial N) \times I$ were merely smooth, we could make it a product by 4.2. Even if only $\partial N \times 0$ were smooth, we could make $(\partial N) \times I$ smooth by 7.4. But we cannot make $\partial N \times 0$ smooth without moving $M \times 0$ — contrary to the requirements of 10.1.

In any case there is a PD isotopy F_t of $(M\times I)_\alpha$ fixed on $(M-N)\times I$, such that $\beta = F_1^*\alpha$ is a product $\theta \times 1$ in a neighborhood of $N \times I$. Put $f_t = F_t|M \times 0$. Then $f_t : M \to M_\omega$ is a PD isotopy rel $M - N$, where $\omega = \alpha|M \times 0$. Put

$$G_t = F_t \circ (f_t^{-1} \times 1_I) : M \times I \to M \times I .$$

Then G_t is fixed on $M \times 0 \cup (M-N) \times I$, and there are diffeomorphisms $G_1 : \omega \times \iota \xrightarrow{f_1^{-1} \times 1} \theta \times \iota \xrightarrow{F_1} \alpha$ in a neighborhood of $N \times I$. Hence $G_1^* \alpha$ is a product in a neighborhood of $N \times I$.

The trouble is that G_t need not be a PD isotopy. By 13.9 however, we can approximate G_t by a PD isotopy $H_t : M \times I \to (M \times I)_\alpha$ such that $H_t = G_t$ on $M \times 0 \cup (M-N) \times I$, for all $t \in I$, and also $H_0 = G_0$, $H_1 = G_1$. This will prove 10.1.

We proceed to fill in some of the details of this argument. Let $N_0 \subset \text{int } N$ be a compact PL submanifold of $K - U$.

10.2 LEMMA. *Given* $\varepsilon > 0$ *there is a* PD *isotopy* $F_t : M \times I \to (M \times I)_\alpha$ *such that*:

(a) $F_1((\partial N_0) \times I$ *is a smooth submanifold*;

(b) *the induced smoothing on* $(\partial N_0) \times I$ *is a product*

(c) F_t *is a product isotopy* $f_t \times 1$ *in a closed neighborhood* $U_0 \times I$ *of* $K \cap \partial N_0$ *in* int N.

(d) F_t *is fixed on a neighborhood of* $(M - \text{int } N) \times I$

(e) $|\pi_2 F - \pi_2| < \varepsilon$.

Proof. By uniqueness of collaring we may assume α is a product in a neighborhood of $N_0 \times I$. By 7.4 there is a small PD isotopy of $(M \times 0, \alpha_0)$ taking ∂N_0 onto a smooth manifold. Extend this isotopy to a small isotopy u_t of $(M \times I)_\alpha$ in such a way that u_t is a product in a neighborhood of $N_0 \times 0 \cup (K \cap \partial N_0) \times I$, and u_t is fixed in a neighborhood of $(M - \text{int } N) \times I$. Let $\beta = u_1 \alpha \in S(M \times I)$. Then $(\partial N_0 \times 0) \cup [(K \cap \partial N_0) \times I] = X$ has a neighborhood in $(\partial N_0) \times I$ which is a smooth submanifold of β; and the smoothing is a product in a neighborhood of $(K \cap \partial N_0) \times I$. By 7.4 there is a small PD isotopy v of $(M \times I)_\beta$ taking $(\partial N_0) \times I$ onto a smooth submanifold of β which is fixed in a neighborhood of X and on a neighborhood of $(M - \text{int } N) \times I$; in particular it is fixed on $M \times 0$. Let

$\gamma = v_1^*\beta$; then $(\partial N_0) \times I$ is a smooth submanifold of γ, and the smoothing is a product in a neighborhood of $(K \cap \partial N_0) \times I$. Observe that the PD isotopy w_t of $(M \times I)_\alpha$,

$$w_t = \begin{cases} u_{2t}, 0 \leq t \leq \frac{1}{2} \\ u_1\, v_{2t-1}, \frac{1}{2} \leq t \leq 1 \end{cases}$$

is a product in a neighborhood of $(K \cap \partial N_0) \times I$. Hence, replacing α by $w_1^*\alpha$, it suffices to prove 10.2 under the extra assumption: $(\partial N_0) \times I$ is a smooth submanifold of $(M \times I)_\alpha$ and the induced smoothing ω of $(\partial N_0) \times I$ is a product in a neighborhood of $(K \cap \partial N_0) \times I$.

Given $\delta > 0$, by 4.2 there is a PD isotopy g_t of $[(\partial N_0) \times I]_\omega$ such that $g_1^*\omega$ is a product smoothing, g_t is fixed in a neighborhood of $(K \cap \partial N_0) \times I$, and $|\pi_2\, g_t - \pi_2| < \delta$. This isotopy extends to the required isotopy F_t of $(M \times I)_\alpha$ if δ is sufficiently small, proving 10.2.

To prove 10.1, define a PD homeomorphism $g : (M \times I) \to (M \times I)_\alpha \times I$ by $g(p, t) = (F_t(p), t)$ where F is as in 10.2. Set $F_t(x, 0) = (u_t(x), 0)$, let $\alpha_0 = \alpha | M \times 0$, and define a PD homeomorphism $u_t \times 1 \times 1 = f : M \times I \times I \to M_{\alpha_0} \times I \times I$ where $\alpha_0 = \alpha | M \times 0$.

We may assume $F_t = F_0$ for $0 \leq t \leq \frac{1}{3}$ and $F_t = F_1$ for $\frac{2}{3} \leq t \leq 1$. Then $gf^{-1} : M_{\alpha_0} \times I \times I \to (M \times I)_\alpha \times I$ is a diffeomorphism on a neighborhood of $A = M \times I \times 0 \cup [M \times 0 \cup (M-N) \times I] \times I \cup (N \times I) \times 1$. Moreover $\pi_3 g = \pi_3 f = \pi_3 : M \times I \times I \to I$. By 13.9 there is a PD homeomorphism

$$h : M \times I \times I \to (M \times I)_\alpha \times I$$

agreeing with gf^{-1} in a neighborhood of A, preserving π_3, and approximating gf^{-1} as closely as desired. Put $h(x, s, t) = (H_t(x, s), t)$. Then $H_t : M \times I \to (M \times I)_\alpha$ satisfies 10.1.

§11. *Proof of the main theorem*

Let K, N, M, α, ε be as in 4.1. Put $m = \dim M$.

Let $\delta : M \times I \to R_+$ be such that if $u_0, \ldots, u_m$ are δ-maps of $M \times I$ then the composition $u_m \circ \ldots \circ u_1$ is an ε-map.

Give M a triangulation such that

(1) $\operatorname{diam} \sigma < \frac{1}{2} \min \{\delta(x) : x \in \sigma\} = \frac{1}{2} \delta_\sigma$,

(2) if $\sigma \cap K \cap \operatorname{bd} N \neq \emptyset$ then $\sigma \subset U$.

(3) if $\sigma \cap (K - u) \neq \emptyset$ then $\sigma \subset \operatorname{int} N$.

Let M_i denote the i-skeleton of the triangulation.

Let L be the union of these simplices meeting $K - U$. Then $L \subset \operatorname{int} N$. Let L_i denote the union of the i-simplices in L. We shall prove the following lemma by induction on $i \in \{-1, 0, \ldots, m\}$; this will suffice to prove 4.1.

11.1 LEMMA. *Suppose* a *is a product in a neighborhood* $U_{i-1} \times I$ *of* $L_{i-1} \times I$. *There exists a* PD *isotopy* $f_t : M \times I \to (M \times I)_a$ *such that*

(a) $f_1^* a$ *is a product in a neighborhood of* $L_i \times 1$

(b) f_t *is fixed on* $M \times 0 \cup (N - N) \times I$

(c) f_t *is a* δ*-isotopy.*

Proof. If $i = -1$ the lemma is trivial; suppose $i \geq 0$, and assume the lemma is true for smaller values of i.

For each i-simplex σ of L we choose a closed PL m-cell $N_\sigma \subset M$ such that:

(4) $\sigma - U_{i-1} \subset \operatorname{int} N_\sigma$

(5) $N_\sigma \cap N_\tau = \emptyset$ if $\sigma = \tau$

(6) $N_\sigma \cap L_{i-1} = \emptyset$

(7) $\operatorname{diam} N_\sigma < \frac{1}{2} \delta_\sigma$ (see (1));

(8) Put $K_\sigma = \sigma \cap N_\sigma$. Then $K_\sigma \cap \partial N_\sigma \subset U_{i-1}$.

Apply Theorem 10.1 to $K_\sigma \subset N_\sigma \subset M$: there exists a PD isotopy $f_{\sigma t} : M \times I \to (M \times I)_a$ such that:

(9) $f_{\sigma t}$ is fixed on $(M - N_\sigma) \times I \cup M \times 0$;

(10) $(f_{\sigma t})^* \alpha$ is a product in a neighborhood of K_σ;

(11) $|\pi_2 f_{\sigma t}(x,s) - s| < \frac{1}{2}\delta_\sigma$

(12) $d(\pi_2 f_{\sigma t}(x,s) - x) < \operatorname{diam} N_\sigma$.

Because of (8) and the hypothesis that $\alpha|U_{i-1} \times I$ is a product, (10) and (9) imply

(13) $(f_{\sigma t})^* \alpha$ is a product on a neighborhood of σ.

From (7), (12) and (11) we have

(14) $f_{\sigma t}$ is a δ-isotopy.

Since the N_σ are disjoint, by (9) we may define a PD isotopy $f_t : M \times I \to (M \times I)_\alpha$ by setting

$$f_t = f_{\sigma t} \text{ on } N_\sigma \times I$$

$$f_t = 1 \text{ on } (M - U_\sigma N_\sigma) \times I .$$

Then f_t satisfies the lemma. The proof of 4.1 is complete.

§12. *Smoothings of* $M \times R^n$

In this section M is a PL manifold.

12.1 THEOREM. *Let* α *be a smoothing of* $M \times R$. *Let* $N \subset M \times R$ *and* $K \times 0 \subset (M \times 0) \cap N$ *be closed sets such that* α *is a product in a neighborhood of* $(K \cap \operatorname{bd} N) \times 0$ *in* $M \times R$. *Given* $\varepsilon : M \times R \to R_+$ *there is a PD* ε*-isotopy* $f_t : M \times R \to (M \times R)_\alpha$ *such that*

(a) $f_1^* \alpha$ *is a product in a neighborhood of* K

(b) f_t *is fixed outside* N

(c) K *has a neighborhood* $W \subset M$ *such that* $f_1(W \times 0)$ *is a smooth submanifold of* α.

Proof. This is a slight refinement of 7.4. An application of 7.4 shows that we may assume that K has a neighborhood $U \subset M$ such that $U \times 0$ is a smooth submanifold of $(M \times R)_\alpha$. Then U has both a smooth collar and a PL collar in $U \times R$; and these can be chosen to coincide where α is a product. The required isotopy is one which pushes the PL collar onto the smooth collar near $K \times 0$ keeping fixed the complement of N, as in 13.10. Note that (c) follows from (a).

The following result with $M = \{0\}$ shows that R^n has a unique smoothing up to isotopy and hence up to diffeomorphism.

12.2 THEOREM. *Let* $K \subset N \subset M$ *be closed sets and* α *a smoothing of* $M \times R^n$ *which is a product in a neighborhood* E *of* $(K \cap \mathrm{bd}\, N) \times R^n$. *Given* $\varepsilon : (M \times R^n) \to R_+$ *there exists a smoothing* β *of* $M \times R^n$ *such that:*

(a) β *is a product in a neighborhood* E' *of* $K \times R^n$

(b) $\beta \cong_\varepsilon \alpha$ rel $(M - N) \times R^n$.

Moreover if E *is a uniform neighborhood* (*i.e.*, $E = E_0 \times R^n$) *then* E' *in* (a) *is a uniform neighborhood.*

Proof. Since $M \times R^n = (M \times R^{n-1}) \times R$, it suffices, by induction on n, to prove the theorem for $n = 1$.

By 12.1 we may assume that for each $n \in Z$ there is a neighborhood N_n of K in M such that $N_n \times n \subset (M \times R)_\alpha$ is a smooth manifold.

Moreover we may assume $\cap_{n \in Z} N_n$ contain a neighborhood W of K in N. By the main theorem applied to $((N_0 \cap N_1 \times [0,1])_\alpha$ we can find a neighborhood $N'_1 \subset N_0 \cap N_1$ of W and an $\varepsilon/4$-isotopy f_t^1 of $(M \times R)_\alpha$ such that f_t^1 is fixed on $M \times \langle -\infty, 0] \cup M \times [\frac{3}{2}, \infty\rangle \cup (M - N) \times R$ and such that $f_1^{*}\alpha$ is a product on $M'_1 \times [0,1]$.

Similarly, if $\alpha_1 = f_1^{1*}\alpha$, there is an $\varepsilon/8$-isotopy f_t^2 of $(M \times R)_\alpha$ fixed on

$$M \times \langle -\infty, 1] \cup M \times [\tfrac{5}{2}, \infty\rangle \cup (M - N) \times R\ ,$$

such that ${f_1^2}^* \alpha_1 = \alpha_2$ is a product on $N_2^1 \times [1,2]$, for some neighborhood $N'_2 \subset N'_1 \cap N_2$ of W. This makes α_2 a product on $N'_2 \times [0,2]$. In this way we find by recursion a sequence of isotopies $g_t^1 = f'_t$, $g_t^2 = f'_1 \circ f_t^2$, $g_t^3 = f_1^1 \circ f_1^2 \circ f_t^3, \ldots,$ of $(M \times R)_\alpha$ such that if we put $h_t = \lim_{n\to\infty} g_t^n$, then $h_t : M \times R \to (M \times R)_\alpha$ is a PD $\varepsilon/2$-isotopy fixed on $M \times \langle -\infty, 0] \cup (M-N) \times R$ such that h_1^* is a product on a neighborhood of $K \times R$. The same construction is applied to $M \times [-1,0]$, $M \times [-2,-1]$, etc. The last statement of the proof follows from the construction.

A version of the following result is called the "product smoothing theorem" in Hirsch [2]. Let $\rho^n \in S(R^n)$ be the standard smoothing.

12.3 THEOREM. *Let* $A \subset M$ *be a closed set. Let* α *be a smoothing of* $M \times R^N$ *which is a product in a neighborhood* E *of* $A \times R^n$. *There is a smoothing* β *of* M *such that* $\alpha \cong \beta \times \rho^n$ rel *a neighborhood* E' *of* $A \times R^n$, *and* E' *is uniform if* E *is uniform. Moreover* α *is unique up to isotopy rel a neighborhood of* A.

Proof. The existence of β follows from 12.2. If also $\gamma \in S(M)$ is such that $\alpha \cong \gamma \times \rho^n$, let

$$f_t : M \times R^n \to M_\beta \times R^n, \quad 0 \leq t \leq 1$$

be an PD isotopy from $\gamma \times \rho^n$ to $\alpha \times \rho^n$, fixed on a neighborhood of $A \times R^n$.

We may assume $f_t = f_0 = 1$ for t near 0 and $f_t = f_1$ for t near 1. Define

$$F : M \times R^n \times I \to M \times R^n \times I$$

by

$$F(x,s,t) = (f_t(x), x, t) \ .$$

Let $\delta \in S(M \times R^n \times I)$ be the smoothing $F^*(\beta \times \delta^n \times \iota)$ induced by F. Observe that $M \times 0 \times I \subset (M \times R^n \times I)_\delta$ is a smooth submanifold of

$M \times 0 \times 0 \cup M \times 0 \times 1 \cup A \times 0 \times I$ and the induced smoothings of $M \times 0 \times 0$ and $M \times 0 \times 1$ are just β and γ while in a neighborhood of $A \times 0 \times I$, δ is a product. By 12.1 we may push $M \times 0 \times I$ onto a smooth submanifold of $(M \times R^n \times I)_\delta$ keeping a neighborhood of $A \times I$ fixed. This gives a smoothing λ of $M \times 0 \times I$ extending β and γ which is a product in a neighborhood of $A \times I$. By the main theorem (4.1) λ is isotopic to $\beta \times \iota$ rel a neighborhood of $A \times I$. Restricting the isotopy to $M \times 1$ proves $\gamma \cong \beta$ rel a neighborhood of $A \times I$.

§13. *Appendix*

Smooth triangulation

We need sharper forms of Whitehead's extension and approximation theorems than those stated in [22] or [16]. The original proofs also prove the stronger theorems.

Let C be a finite complex in R^q. Our first observation concerns the topology on the vector space $\mathrm{Simp}(C, R^p)$ of simplicially smooth maps $C \to R^p$. We define, for $f \in \mathrm{Simp}(C, R^p)$,

$$|f| = \sup\{|f(x)| : x \in |C|\}$$

$$\mathrm{Lip}(f) = \sup\{|f(x) - f(y)|/|x-y| : x, y \in |C|, x \neq y\}$$

$$\|f\|_{\mathrm{Lip}} = \max\{|f|, \mathrm{Lip}(f)\} .$$

Then $\|f\|_{\mathrm{Lip}}$ is always finite and defines a norm on $\mathrm{Simp}(C, R^p)$, the *Lipschitz norm.* It is easy to see that the topology induced by the Lipschitz norm is the same as the C^1 topology of Munkres [16].

A map $f \in \mathrm{Simp}(C, R^p)$ is *nondegenerate* if $|C|$ is covered by open sets $U \subset |C|$ such that $f|U|$ is invertible and $\mathrm{Lip}[(f|U)^{-1}] < \infty$. The set of nondegenerate maps is open in $\mathrm{Simp}(C, R^p)$.

Let $P \subset R^q$ be a compact polyhedron. We denote by $PD(P, R^q)$ the union of the sets $\mathrm{Simp}(C, R^q)$ where C runs through all (rectilinear) triangulations of P. We give $PD(P, R^q)$ the Lipschitz norm.

Given $f \in PD(P, R^q)$ and a triangulation C of P, Whitehead defines a simplicially affine map $\mathcal{L}_C f : C \to R^p$ by setting $\mathcal{L}_C f(v) = f(v)$ for every vertex $v \in C$. He proves that if f is nondegenerate, then for any $\varepsilon > 0$ and any triangulation C_0 of P there exists a refinement C of C_0 such that

$$\|f - \mathcal{L}_C f\|_{Lip} < \varepsilon .$$

Our second remark concerning Whitehead's theorems is that if $\phi : R^p \to R^s$ is an affine map such that $\phi f : C_0 \to R^s$ is simplicially affine, then $\phi \circ \mathcal{L}_C f = \phi f$ for every subdivision C of C_0. We have proved:

13.1 THEOREM. *Let* $P \subset R^q$ *be a compact polyhedron and* $f : P \to R^p$ *a nondegenerate* PD *map. Let* $\phi : R^p \to R^s$ *be affine and suppose* $\phi f : P \to R^s$ *is* PL. *Let* $P_0 \subset P$ *be a compact polyhedron such that* $f|P_0$ *is* PL. *Given* $\varepsilon > 0$ *there exists a* PL *map* $g : P \to R^p$ *such that*

(a) $\|f - g\|_{Lip} < \varepsilon$

(b) $\phi g = \phi f$

(c) $g|P_0 = f|P_0$.

Another of Whitehead's basic tools is an extension theorem. Let $\Delta \subset R^q$ be a simplex with barycenter b, and $f_0 : \Delta \to R^p$ a smooth map. Let $g_0 : \partial\Delta \to R^p$ be an approximation to $f|\partial\Delta$. We extend g_0 to a map $g : \Delta \to R^p$ as follows. Every point $y \in \Delta$ can be expressed uniquely as a convex combination

$$(1) \quad y = (1-t)b + ty', \quad 0 \le t \le 1, \quad y' \in |\partial\Delta| .$$

Let $y'' = \frac{1}{2}b + \frac{1}{2}y'$ be the midpoint of $[b, y']$. If $t \ge \frac{1}{2}$ in (1), put $y = (1-s)y'' + sy'$. Define

$$(2) \quad g(y) = \begin{cases} f(y) & \text{if } 0 \le t \le \frac{1}{2} \\ (1-s)g(y'') + sg_0(y') & \text{if } \frac{1}{2} \le t \le 1. \end{cases}$$

Thus $g = f$ on $[b, y'']$, $g(y') = g_0(y')$ and g is affine on $[y'', y']$ if $t \geq \frac{1}{2}$. Whitehead shows that g_0 is simplicially smooth in a particular subdivision of Δ, and that $\|g-f\|_{\mathrm{Lip}} \to 0$ as $\|g_0 - f|\partial\Delta\|_{\mathrm{Lip}} \to 0$. Working simplex by simplex, he then proves an extension theorem for PD maps of polyhedra. Our addition to this theorem is simply the observation that if $\phi : \mathbf{R}^p \to \mathbf{R}^s$ is an affine map such that $\phi f : \Delta \to \mathbf{R}^s$ is affine and $\phi g_0 = \phi f|\partial\Delta$, then $\phi g = \phi f$, where g is defined in (2). Therefore we have proved:

13.2 THEOREM. *Let* $P \subset \mathbf{R}^q$ *and* $P_0 \subset P$ *be compact polyhedra,* $f : P \to \mathbf{R}^p$ *a* PD *map and* $\phi : \mathbf{R}^p \to \mathbf{R}^s$ *and affine map such that* $\phi f : P \to \mathbf{R}^s$ *is* PL. *Given* $\varepsilon > 0$ *there exists* $\delta > 0$ *with the following property. If* $g_0 : P_0 \to \mathbf{R}^p$ *is a* PD *map satisfying*

$$\|g_0 - f|P_0\|_{\mathrm{Lip}} < \delta$$

and

$$\phi g_0 = \phi f|P_0$$

there exists a PD *map* $g : P \to \mathbf{R}^p$ *such that*:

(a) $g|P_0 = g_0$

(b) $\|g-f\|_{\mathrm{Lip}} < \varepsilon$

(c) $\phi g = \phi f$.

We combine 13.1 and 13.2 to get a strongly relative theorem for PL Lipschitz approximations to PD maps into Euclidean space:

13.3 THEOREM. *Let* $P \subset \mathbf{R}^q$ *and* $A \subset P$, $B \subset P$ *be compact polyhedra, and* $f : P \to \mathbf{R}^p$ *a nondegenerate* PD *map such that* $f|A \cap B$ *is* PL. *Let* $\phi : \mathbf{R}^p \to \mathbf{R}^s$ *be an affine map such that* $\phi f : P \to \mathbf{R}^s$ *is* PL. *Given* $\varepsilon > 0$, *there exists a* PD *map* $g : P \to \mathbf{R}^p$ *such that*:

(a) $g|A$ *is* PL

(b) $g|B = f|B$

(c) $\|g-f\|_{\mathrm{Lip}} < \varepsilon$

(d) $\phi g = \phi f$.

Proof. Let $g_1 : A \to R^p$ be a PL approximation to $f|A$, agreeing with f on $A \cap B$ such that $\phi g_1 = \phi f|A$ by 13.1. Let $g_2 : A \cup B \to R^p$ be the map equal to g_1 on A and f on B. If $\|g_1 - f|A\|_{Lip}$ is small enough, then by 13.2 g_2 extends to the required map $g : P \to R^p$.

Our next task is to extend these theorems to manifolds. If $M \subset R^p$ is a smooth manifold, we consider PD maps $P \to R^p$ whose image lies in M. The theorems above do not apply directly, since the approximations would not have images in M. Instead we take advantage of the strongly relative form of the theorems and use them repeatedly inside coordinate charts of M. (This is Whitehead's procedure.) In this way, given two PD homeomorphisms $f_i : P_i \to M$ $(i = 0, 1)$, we approximate them by $g_i : P_i \to M$ so that $g_1^{-1} g_0 \to P_1$ is a PL homeomorphism. Going one step further, we keep f_1 fixed and replace g_0 by $h_0 = f_1 g_1^{-1} g_0$ which is PD because $g_1^{-1} g_0$ is PL. Then $f_1^{-1} h_0$ is PL. Moreover $\|h_0 - f_0\|_{Lip} \to 0$ as $\|g_0 - f_0\|_{Lip} \to 0$ and $\|g_1 - f_1\|_{Lip} \to 0$ since

$$\|h_0 - f_0\|_{Lip} \leq \|h_0 - g_0\|_{Lip} + \|g_0 - f_0\|_{Lip}$$

and

$$h_0 - g_0 = (f_1 - g_1) g_1^{-1} g_0, \text{ etc.}$$

In order to state the sharpened form of the uniqueness theorem for smooth triangulations, we make the following definition. A map $\phi : M \to R^s$ is *pseudo-affine* if M is a smooth manifold having an atlas of charts $\xi : U \to R^m$ such that each map

$$\phi \xi^{-1} : \xi(U) \to R^s$$

is the restriction to $\xi(U)$ of an affine map $R^m \to R^s$. The atlas is not required to be maximal. For example M might be a product $N \times R^s$ and $\phi = \pi_2 : N \times R^s \to R^s$.

The following theorem can be proved as sketched above:

13.4 THEOREM. *Let* $M \subset R^p$ *be a smooth manifold and* $P_i \subset R^q$ *polyhedra,* $i = 0, 1$. *Let* $f_0 : P_0 \to M$ *be a nondegenerate* PD *map and* $f_1 : P_1 \to M$ *a* PD *homeomorphism (necessarily nondegenerate). Let* $A \subset P_0$ *be a closed polyhedron such that* $P_0 - A$ *has compact closure and* $f_1^{-1} f_0 | A \to P_1$ *is* PL. *Let* $\phi : M \to R^s$ *be a pseudo-affine map such that* $\phi f_i : P_i \to R^s$ *is* PL, $i = 0, 1$. *Given* $\varepsilon > 0$ *there exists a nondegenerate* PD *map* $h_0 : P_0 \to M$ *such that*

(a) $f_1^{-1} h_0 : P_0 \to P_1$ *is* PL

(b) $\|h_0 - f_0\|_{Lip} < \varepsilon$

(c) $h_0 | A = f_0 | A$

(d) $\phi h_0 = \phi f_0$.

For simplicity we have stated 13.4 in a relative form, leaving a strongly relative theorem as an exercise.

The extension theorem 13.2 has the following globalization.

13.5 THEOREM. *Let* $M \subset R^p$ *be a smooth submanifold,* $P \subset R^q$ *a polyhedron and* $P_0 \subset P$ *a compact polyhedron. Let* $f : P \to M$ *be a nondegenerate* PD *map and* $\phi : M \to R^s$ *a pseudo-affine map such that* $\phi f : P \to R^s$ *is* PL. *Given* $\varepsilon > 0$ *there exists* $\delta > 0$ *with the following property. Let* $g_0 : P_0 \to M$ *be a* PD *map such that*

$$\|g_0 - f | P_0\|_{Lip} < \delta$$

and

$$\phi g_0 = \phi f | P_0 .$$

Then there exists a nondegenerate PD *map* $g : P \to M$ *such that*

(a) $g | P_0 = g_0$

(b) $\|g - f\|_{Lip} < \varepsilon$

(c) $\phi g = \phi f$.

The proof is left to the reader.

The globalization of 13.3 to manifolds is the following strongly relative PL approximation theorem. Again the proof is left to the reader.

13.6 THEOREM. *Let* $M \subset \mathbf{R}^p$ *be a smooth manifold. Let* $P \subset \mathbf{R}^q$, $Q \subset \mathbf{R}^q$, $A \subset Q$ *and* $B \subset Q$ *be polyhedra, with* B *closed in* Q *and* A *compact. Let* $h : P \to M$ *be a* PD *homeomorphism,* $f : Q \to M$ *a nondegenerate* PD *map and suppose* $h^{-1}f : A \cap B \to P$ *is* PL. *Let* $\phi : M \to \mathbf{R}^s$ *be a pseudo-affine map such that both* ϕh *and* ϕf *are* PL. *Given* $\varepsilon > 0$ *there exists a nondegenerate* PD *map* $g : Q \to M$ *such that:*

(a) $h^{-1}g|A : A \to P$ *is* PL

(b) $g|B = f|B$

(c) $\|g - f\|_{Lip} < \varepsilon$

(d) $\phi g = \phi f$.

Compositions of Lipschitz maps

We have implicitly used the well known fact if ϕ is a Lipschitz map and u, v maps such that $\phi \circ u$ and $\phi \circ u - \phi \circ v$ make sense, then $\|\phi u - \phi v\|_0 \to 0$ as $\|u - v\| \to 0$; and also $\mathrm{Lip}(\phi u) \leq \mathrm{Lip}(\phi)\,\mathrm{Lip}(u)$. It is not in general true that $\mathrm{Lip}(\phi u - \phi v) \to 0$ as $\|u - v\|_{Lip} \to 0$. The following lemma states that this is true if ϕ is C^1.

13.7 LEMMA. *Let* $X \subset \mathbf{R}^p$ *be a compact subset and* $M \subset \mathbf{R}^q$ *a smooth submanifold. Let* $u : X \to M$ *be a Lipschitz map and* $\phi : M \to \mathbf{R}^q$ *a* C^1 *map. Given* $\varepsilon > 0$ *there exists* $\delta > 0$ *such that if* $v : X \to M$ *satisfies* $\|v - u\|_{Lip} < \delta$, *then* $\|\phi u - \phi v\|_{Lip} < \varepsilon$.

Proof. By compactness of X it suffices to prove: for any $x \in M$ there exists $\delta_x > 0$ and a neighborhood U_x of x in X such that $\|\phi v - \phi u|U_x\|_{Lip} < \varepsilon$ for any $v : U_x \to M$ such that $\|v - u|U_x\|_{Lip} < \delta_x$. By choosing local coordinates in M we may assume $M = \mathbf{R}^p$.

Given $x \in X$ put $y = u(x) \in \mathbf{R}^p$. Let T denote the linear map $D\phi_y : \mathbf{R}^p \to \mathbf{R}^q$ and put $\phi = T + r$. Put $v = u + h$. Then

$$\begin{aligned}\phi v - \phi u &= (T+r)(u+h) - (T+r)u \\ &= Th + r(u+h) - ru .\end{aligned}$$

Hence $\mathrm{Lip}(\phi v - \phi u) \le \|T\|\ \mathrm{Lip}(h) + \mathrm{Lip}(r)(2\ \mathrm{Lip}(u) + \mathrm{Lip}(h))$. Since $\mathrm{Lip}(r|V)$ can be made arbitrarily small by choosing a suitably small neighborhood V of y in R^p we can make $\mathrm{Lip}(\phi v - \phi u)$ as small as desired by choosing U_x and $\|h\|_{\mathrm{Lip}}$ sufficiently small.

13.8 COROLLARY. *Let* $N, M \subset R^p$ *be compact smooth manifolds and* $\phi : M \to N$ *a diffeomorphism. Given* $\varepsilon > 0$ *there exists* $\delta > 0$ *such that if* $\Psi : M \to N$ *satisfies* $\|\Psi - \phi\|_{\mathrm{Lip}} < \delta$ *then* Ψ *is a homeomorphism and* $\|\psi^{-1} - \phi^{-1}\|_{\mathrm{Lip}} < \varepsilon$.

Proof. The inverse function theorem for Lipschitz maps [7] guarantees that ψ is invertible and $|\mathrm{Lip}(\psi^{-1}) - \mathrm{Lip}(\phi^{-1})|$ small if $\|\psi - \phi\|_{\mathrm{Lip}}$ is sufficiently small. (In fact this is true provided only that ϕ and ϕ^{-1} are Lipschitz.) Writing

$$\phi^{-1} - \psi^{-1} = (\phi^{-1}\psi - \phi^{-1}\phi)\psi^{-1}$$

we have

$$\mathrm{Lip}(\phi^{-1} - \psi^{-1}) \le \mathrm{Lip}(\phi^{-1}\psi - \phi^{-1}\phi)\ \mathrm{Lip}(\psi^{-1})$$

and 13.8 follows from 13.7.

Inverses and compositions of PD *homeomorphisms*

PD maps do not form a category. In general PD maps cannot be composed, nor can PD homeomorphisms be inverted, to produce PD maps. This antisocial behavior causes endless technical difficulties. Thanks to Whitehead's theorems, however, these can usually be overcome. The following application of his approximation theorems is used in Section 10 produce a PD isotopy approximating $f_t^{-1} g_t$ where f_t and g_t are PD isotopies. (It is easier to approximate something of the form $f_t g_t$; we leave the formulation and proof of the appropriate theorem to the reader as an exercise.)

The result below, while sufficient for our purposes, seems to require a certain smoothness hypothesis. It would be more satisfactory to have a statement involving only PD maps.

13.9 LEMMA. *Let* P, Q, M, *be* PL *manifolds; let* α *and* β *be smoothings of* P *and* Q *respectively. Let* $f: M \to P_\alpha$, $g: M \to Q_\beta$ *be* PD *homeomorphisms. Suppose* $A \subset P$ *is a closed set such that* $gf^{-1}: P_\alpha \to Q_\beta$ *is a diffeomorphism in a neighborhood* U *of* A, *and* $P - U$ *is compact. Given* $\varepsilon > 0$ *there exists a* PD *homeomorphism* $h: P \to Q_\beta$ *such that*

(a) $h = gf^{-1}$ *in a neighborhood of* A,

(b) $|h - gf^{-1}| < \varepsilon$.

Moreover if $\phi: P \to R^s$, $\psi: M \to R^s$, $\theta: Q \to R^s$ *are* PL *maps such that* $\phi: P_\alpha \to R^s$ *and* $\theta: Q_\beta \to R$ *are pseudo-affine, and* $\phi f = \psi$, $\theta g = \psi$, *then*

(c) $\theta h = \phi$.

Proof. Let n be the dimension of P, M and Q. Let $B_1 \subset B \subset P$ and $C \subset M$ be compact PL submanifolds of dimension n such that

$$P - U \subset \operatorname{int} B_1 \subset \operatorname{int} f(C),$$

$$f(C) \subset \operatorname{int} B$$

$$B \subset P - A.$$

Given $\delta > 0$, by 13.6 there exists a PD homeomorphism $f_0: M \to P_\alpha$ such that:

$$f_0|C \text{ is PL}$$

$$f_0 = f \text{ on } M - f^{-1}B$$

$$\|f - f_0\|_{\mathrm{Lip}} < \delta .$$

(For statements about norms we assume P, M, Q embedded in R^q as polyhedra and P_α, Q_β embedded in R^p as smooth submanifolds.)

Consider the PD map

$$gf_0^{-1}|\partial B_1$$

as an approximation on ∂B_1 to the PD map $gf^{-1}|B - \text{int } B_1 : B - \text{int } B_1 \to Q_\beta$, this last map being PD because $B - \text{int } B_1 \subset U$ by (1). By 13.5 if

$$\|(gf_0^{-1} - gf^{-1})|\partial B_1\|_{\text{Lip}}$$

is small enough, there exists a PD map $h_0 : B - \text{int } B_1 \to Q_\beta$ such that $h_0 = gf_0^{-1}$ on ∂B_1, $\theta h_0 = \phi$ on $B - \text{int } B_1$ and $\|h_0 - gf^{-1}\|_{\text{Lip}} < \varepsilon$. Extend h_0 to $h : P \to Q$ by making $h = gf_0^{-1}$ on B_1. This completes the proof of 13.9.

Observe that we lose control of $\text{Lip}(gf_0^{-1})$ outside of U; for this reason the approximation (b) of 13.9 is only C^0.

Uniqueness of collaring

13.10 PROPOSITION. *Let* M *be a* PL *manifold,* $\partial M = \emptyset$ *and* α *a smoothing of* $M \times R_+$. *Let* $N \subset M \times R_+$ *and* $K \times 0 \subset (M \times 0) \cap N$ *be closed sets. Let* $W \subset M \times R$ *be an open set containing* $M \times 0 \cup N$ *and let*

$$f : W \to (M \times R_+)_\alpha$$

be a PD *embedding. Suppose*

(a) $f|M \times 0 = 1$

(b) $f = 1$ *in a neighborhood* $U \subset W$ *of* $K \times 0 \cap \text{bd}\, N$.

Given $\varepsilon : W \to R_+$ *there is a* PD ε*-isotopy* $f_t : W \to (M \times R)_\alpha$ *of* $f = f_0$ *such that*

(c) $f_t = f$ *outside* N *and on* $M \times 0$, $0 \le t \le 1$

(d) $f_1 = 1$ *in a neighborhood of* $K \times 0$ *in* W.

Proof. We may assume $N \subset W \cap f(W)$. Since the proposition is strongly relative in form (see Section 2) we shall prove it only for the case K compact, leaving the easy globalization to the reader. In this case we may assume $N = N_0 \times [0, q]$, $q > 0$, where $N_0 \subset M$ is a compact set containing K; and $U = U_0 \times [0, q]$ where $U_0 \subset M$ is an open neighborhood of $K \cap \text{bd}\, N_0$. For simplicity of notation assume $q = 1$.

By 13.7 we can make f PL in a neighborhood of $K \times 0$, keeping f fixed on $W - N$. Therefore we assume f is PL. Henceforth we ignore smoothings and work entirely in the PL category.

Choose $0 < b < c < 1$ and a compact neighborhood $N_1 \subset \operatorname{int} N_0$ of $K_0 = K - U_0$ such that

$$f(N_1 \times [0,0]) \subset N_0 \times [0,1\rangle$$

$$f^{-1}(N_1 \times [0,0]) \subset N_0 \times [0,1\rangle$$

$$N_1 \times [0,2b] \subset f(N_0 \times [0,0\rangle) \ .$$

Extend f to $F : W \cup (M \times \langle -\infty, 0]) \to M \times \mathbf{R}$, setting $F|M \times \langle -\infty, 0] = 1$.

The reader can supply a PL isotopy ϕ_t of $M \times \mathbf{R}$ such that if $(x,y) \in N_1 \times \langle -\infty, b]$ then $\phi_1(x,y) = (x, y-b)$, while $\phi_t = 1$ on $(M - N_\delta) \times \mathbf{R}_+ \cup N_0 \times [c, \infty\rangle$.

The isotopy $f_t = \phi_t^{-1} F \phi_t | W : W \to M \times \mathbf{R}_+$ has the required properties.

The Alexander trick

The following beautiful idea, due to Alexander [1] is what makes the theory of PD isotopies apparently more powerful than that of prismatic isotopies.

Alexander worked in what we would call the topological category. The proof below is essentially his proof given a geometrical interpretation.

13.11 THEOREM (Alexander). *Let* E *be a* PL *n-cell and* $f : E \to E$ *a* PL *homeomorphism fixed on* ∂E. *Then* f *is* PL *isotopic to* 1 rel ∂E.

Proof. Take E as a convex polyhedron in $\mathbf{R}^n$ with 0 as an interior point. Define F to be the homeomorphism of $X = E \times 1 \cup (\partial E) \times I$ which is f on $E \times 1$ and the identity on $(\partial E) \times I$. Consider $E \times I \subset \mathbf{R}^n \times \mathbf{R}$ to be the cone from $(0,0)$ on X. Let $G : E \times I \to E \times I$ be the cone on F, i.e., the radial extension of F. Since G preserves $\pi_2 : E \times I \to I$, we can write $G(x,t) = (f_t(x), t)$. Then $f_1 = f$ and $f_0 = 1_E$; hence f_t is a PL isotopy from 1_E to f.

BIBLIOGRAPHY

[1]. Alexander, J., On deformations of the n-cell. Proc. Nat. Acad. Sc. 9 (1923), pp. 406-407.

[2]. Cairns, S., Homeomorphisms between topological manifolds and analytic manifolds. Annals of Math. 41 (1940), pp. 796-808.

[3]. Hirsch, M. W., Obstructions to smoothing manifolds and maps. Bull. Amer. Math. Soc. 69 (1963), pp. 352-356.

[4]. ———, On tangential equivalence of manifolds. Ann. of Math. 83 (1966).

[5]. ——— and Mazur, B., Smoothings of piecewise linear manifolds. Cambridge Univ. 1964 (mimeo).

[6]. ——— and Milnor, J., Some curious involutions of spheres. Bull. Amer. Math. Soc.

[7]. ——— and Pugh, C., Stable manifolds for hyperbolic sets, Proceedings of *Symposia in Pure Mathematics* XIV, Providence 1970.

[8]. Kervaire, M., A manifold which does not admit any differentiable structure. Comm. Math. Helv. 34 (1960), pp. 304-312.

[9]. Kuiper, N., On smoothings of triangulated and combinatorial manifolds. *Differential and Combinatorial topology*. A symposium in honor of Marston Morse. Princeton 1965, pp. 3-22.

[10]. Lashof, R., and Rothenberg, M., Microbundles and smoothing. Topology 3 (1965), pp. 357-380.

[11]. Mazur, B., Séminaire de Topòlogie combinatoire et differentielle. Inst. Hautes Etudes Scien. 1962.

[12]. Munkres, J., Obstruction to the smoothing of piecewise differentiable homeomorphisms. Ann. of Math. 72 (1960), pp. 521-554.

[13]. ———, Obstructions to imposing differentiable structures. Ill. J. Math. 8 (1964), pp. 361-376.

[14]. ———, Concordance is equivalent to smoothability. Topology 5 (1966), pp. 371-389.

[15]. ———, Compatibility of imposed differentiable structures. Ill. J. Math. 12 (1968), pp. 610-615.

[16]. ——————, Elementary Differential Topology. Ann. of Math. Study no. 54, Princeton 1961.

[17]. ——————, Concordance of differential structures: two approaches. Mich. Math. J. 14 (1967), pp. 183-191.

[18]. Rourke,C., and Sanderson, B., Block Bundles. Annals of Math. 87 (1968), pp. 1-28, 255-277, 431-483.

[19]. Smale, S., Generalized Poincare's Conjecture in dimensions greater than 4. Annals Math. 74 (1961), pp. 391-406.

[20]. Sullivan, D., On the Hauptvermutung for manifolds. Bull. Amer. Math. Soc. 63 (1967), pp. 598-600.

[21]. Thom, R., Des variétés triangulées aux variétés différentiables. Proc. Int. Cong. Math. Edinburgh 1958, pp. 248-255.

[22]. Whitehead, J. H. C., On C^1 complexes. Ann. Math. 41 (1940), pp. 809-824.

[23]. ——————, Manifolds with transverse fields in Euclidean space. Ann. Math. 73 (1961), pp. 154-212.

[24]. Zeeman, E. C., and Hudson, J. F. P., On regular neighborhoods. Proc. London Math. Soc. 14 (1964), pp. 714-745.

SMOOTHINGS OF PIECEWISE LINEAR MANIFOLDS II: CLASSIFICATION

Morris W. Hirsch and Barry Mazur

Introduction

Let $\mathcal{S}(M)$ denote the set of concordance classes of smoothings of a PL manifold M. The object of this paper is to reduce the study of $\mathcal{S}(M)$ to tractable, or at least standard, topological problems. This is done in several related ways.

The first approach is based on the fact that if M is smoothable, then the diagonal $M_\Delta \subset M \times M$ has a vector bundle neighborhood. It turns out that $\mathcal{S}(M)$ is naturally isomorphic to the set of stable equivalence classes of such neighborhoods. The proof of this is purely geometrical.

Block bundle theory identifies the set of stable equivalence classes of vector bundle neighborhoods of M_Δ in $M \times M$ with the set of homotopy classes of sections of a fibration $\mathfrak{S}(M)$ over M with fibre PL/0. This fibration is induced from the fibration $BO \to BPL$ by the map $M \to BPL$ that classifies the stable normal block bundle of M_Δ in $M \times M$ (or equivalently, the stable tangent microbundle of M). In this way the study of $\mathcal{S}(M)$ is reduced to the study of sections of the fibre space $\mathfrak{S}(M)$.

For any $\alpha \in \mathcal{S}(M)$ one can establish, by geometric means, an abelian group structure on $\mathcal{S}(M)$ with α equal to the origin.

Forming $\Gamma = PL/O$, the fibre of the fibration (ε) $BPL \to BO$, we may also establish a natural isomorphism

$$(*) \qquad [M, \Gamma] \ \overset{\rightarrow}{\approx}\ \mathcal{S}(M)$$

taking O to α.

By such considerations, the functor $[-,\Gamma]$ is endowed with an abelian group structure. Consequently Γ is endowed with a homotopy abelian and homotopy associative h-space structure.

The fibration (ε) may then be shown to be Γ-principal, and $\mathcal{S}(M)$ corresponds to homotopy classes of its sections.

Going full circle, $\mathcal{S}(M)$ is then a principle homogeneous space over the abelian group $[M,\Gamma]$. Moreover, choosing $\alpha \in \mathcal{S}(M)$, the natural isomorphism (*) above is given by the rule:

$$\begin{aligned} [M,\Gamma] &\to \mathcal{S}(M) \\ \gamma &\mapsto \gamma \cdot \alpha \end{aligned}$$

If M is a *smooth* manifold, let $\mathcal{S}(M)$ denote $\mathcal{S}(M')$ where M' is some PL manifold isomorphic to a smooth triangulation of M. Then there is then a *natural* isomorphism $\mathcal{S}(M) \approx [M,\Gamma]$. In other words, the functor $M \mapsto \mathcal{S}(M)$, from smooth manifolds to sets, is equivalent to the *homotopy* functor $M \mapsto [M,\Gamma]$, from smooth manifolds to abelian groups.

Historical remarks

At the International Congress of Mathematicians in 1958, René Thom [29] suggested that smoothings ought to be classified by sections of a fibration. Subsequently J. Munkres [17; 18] found obstruction theories for certain smoothing problems, and A. Gleason (1959, unpublished) proved that every contractible open PL manifold could be smoothed. These results were further evidence for Thom's idea. Milnor conjectured that $\mathcal{S}(S^n)$ is isomorphic to $\pi_n(PL/O)$, which suggested that the fibre should be PL/O. Moreover, he invented microbundles [15; 16] and showed that a PL manifold M is smoothable if and only if its tangent microbundle has a vector bundle structure. In the meantime the stability theorem $\mathcal{S}(M) \approx \mathcal{S}(M \times \mathcal{R})$ was proved by Hirsch [6] and Mazur-Poenaru [14] and was used to verify Milnor's conjecture.† Another obstruction theory was obtained by Hirsch [6].

† We denote the real number by $\mathcal{R}$ and Euclidean n-space by $\mathcal{R}^n$.

The classification of $\mathcal{S}(M)$ by liftings of $M \to BPL$ over $BO \to BPL$ was proved in Mazur-Poenaru [14]. The subject was much discussed at the topology conference held in Seattle in 1963, where the general outline emerged of a classification of concordance classes of smoothings of M by maps $M \to PL/O$. Lashof and Rothenberg [13] published a partial classification based on these ideas. In 1966 Haefliger [5] and others obtained results on "ambient" smoothings of a PL submanifold of a smooth manifold, in connection with smooth knots. Recently D. White [30] has classified ambient smoothings using semisimplicial complexes and block bundles.

The present paper was begun in 1963 at Cambridge University, following the Seattle conference. A set of mimeographed notes was produced but political turmoil at the University of California interrupted completion of the paper. In the meantime new results in topology were discovered; some of these have been used to simplify the theory of smoothings. In particular the theory of block bundles of Rourke-Sanderson [23], Morlet [17] and Kato [11] makes it possible to avoid the use of semisimplicial complexes (except in Section 6) and the stable PL tubular neighborhood theorem of Lashof-Rothenberg. (This theorem is used implicitly in Section 6 in identifying the classifying spaces for stable PL microbundles and stable block bundles.)

Outline of this paper

The concordance classification of smoothings of a PL manifold M is carried out in stages. First the problem is shown to be stable: the natural map $\mathcal{S}(M) \to \mathcal{S}(M \times \mathcal{R}^n)$ is bijective, where $\mathcal{S}(M)$ is the set of concordance classes of smoothings of M. This was done in Part I. The next step is to extend this to nontrivial vector bundles; there is a bijective map

$$x^! : \mathcal{S}(M) \to \mathcal{S}(E)$$

where $x = (p, E, M)$ is a piecewise differentiable vector bundle. This is carried out in Sections 1 and 2; the main results are 2.6 and its variations 2.7, 2.8 and 2.9.

In Section 3 linearizations of a PL manifold pair (V, M) are studied. A linearization of (V, M) is a PD vector bundle (p, E, M) where $E \subset V$ is a neighborhood of M and $p: E \to M$ is a retraction; the set of equivalence classes of linearizations of (V, M) is called $\mathcal{L}(V, M)$. The direct limit:

$$\cdots \to \mathcal{L}(V \times \mathcal{R}^n, M \times 0) \to \mathcal{L}(V \times \mathcal{R}^{n+1}, M \times 0) \to \cdots$$

is $\mathcal{L}_s(V, M)$, the set of stable linearization classes.

The main results of Section 3 are as follows. Theorem 3.2 says: M *is smoothable if and only if* $\mathcal{L}_s(M \times M, M_\Delta) \neq \phi$, where M_Δ is the diagonal. This is a strengthening of Milnor's theorem that M is smoothable if and only if the tangent microbundle of M has a vector bundle structure. Theorem 3.4 says: *the map*

$$\Phi : \mathcal{L}_s(V, M) \times \mathcal{S}(M) \to \mathcal{S}(V, M)$$

is a complete pairing. Here $\mathcal{S}(V, M)$ is the set of concordance classes of germs of smoothings of neighborhoods of M in V. The map Φ assigns to a linearization x of $(V \times \mathcal{R}^m, M \times 0)$ and a smoothing α of $V \times \mathcal{R}^m$, the smoothing β of M such that $M_\beta \times \mathcal{R}^m$ is concordant to $x^! \alpha$ in a neighborhood of $M \times 0$ in $V \times \mathcal{R}^m$. To say Φ is a complete pairing means that the maps $\alpha \mapsto \Phi(x, \alpha)$ and $x \mapsto \Phi(x, \alpha)$ are bijective when x or α, respectively, is fixed.

The final result of Section 3 is the classification Theorem 3.10. This says that *the map* $\mathrm{Exp}: \mathcal{S}(M) \to \mathcal{L}_s(M \times M, M_\Delta)$ *is bijective*, where Exp assigns to a smoothing α of M the stable linearization of a neighborhood of the diagonal, obtained from the exponential map of a Riemannian metric for α. The proof of 3.10 is based on Lemmas 3.8 and 3.9, which ultimately depend on the smooth tubular neighborhood theorem for $M_\Delta \subset M \times M$. Lemma 3.8 says that if $\alpha, \omega \in \mathcal{S}(M)$ then $(\mathrm{Exp}\, \alpha)^! \omega = (\omega \times \alpha)_\Delta = (\alpha \times \omega)_\Delta$ where $(\quad)_\Delta$ means the M_Δ germ of a smoothing of a neighborhood of the diagonal, and $\mathrm{Exp}: \mathcal{S}(M) \to \mathcal{L}(M \times M, M_\Delta)$ is the unstabilized form of Exp. Lemma 3.9 interprets 3.8 to prove:

$$\Phi(\mathrm{Exp}\,\alpha,\omega) = \Phi(\mathrm{Exp}\,\omega,\alpha)\ .$$

Theorem 3.10 is a formal consequence of this relation and 3.4.

The classification 3.10 is an isomorphism between two sets of geometrical objects: smoothings of M, and stable linearizations of a neighborhood of the diagonal. In Section 4 the study of $\mathcal{L}_s(V,M)$ is reduced to a homotopy question by the theory of block bundles. Therefore: $\mathcal{S}(M)$ *is isomorphic to homotopy classes of liftings of the map* $M \to BPL$ *over* $t: BO \to BPL$. Here $M \to BPL$ classifies the stable tangent block bundle of M, or equivalently, the stable tangent microbundle of M; while $t: BO \to BPL$ is defined by triangulating vector bundles. Pulling back the fibration t over M via $M \to BPL$ gives an induced fibration $\mathcal{E}(M) \to M$ whose fibre is PL/O. Therefore $\mathcal{S}(M) \approx$ homotopy classes of sections of $\mathcal{E}(M)$.

The fibre $\Gamma = PL/O$ turns out to be an h-space which is homotopy associative and homotopy commutative. This is proved in Section 4 using the finiteness of $\mathcal{S}(S^n)$ and a geometrically defined abelian group structure on $\mathcal{S}(M)$ (for smoothable M).

The fibration $\mathcal{E}(M)$ is shown to be Γ-principle and the isomorphism $\mathcal{S}(M) \approx [M,\Gamma]$ obtained.

Obstruction theories for smoothing a manifold, or a PD homeomorphism between smooth manifolds, are derived in Section 5.

In Section 6 an alternative description of h-structure on PL/O is given, relating it to the h-structures on BO and BPL. No homotopy theory is used.

Results used from I

In deriving the geometric classification 3.10, the main result needed from Part I is the $M \times \mathcal{R}^n$ theorem: *the natural map* $\mathcal{S}(M) \to \mathcal{S}(M \times \mathcal{R}^n)$ *is bijective*. This is Theorem 7.8 of I. The stronger $M \times I$ theorem is needed whenever the fact that concordance implies isotopy is used; for the proof in Section 4 (but not Section 6) that PL/O is an h-space, for example, and in developing some of the obstruction theories in Section 5.

§1. *Vector bundles*

(1.1) *A vector bundle atlas* Φ on a map $p: E \to B$ is a set $\Phi = \{(\phi_i, U_i)\}_{i \in \Lambda}$ of homeomorphisms $\phi_i : p^{-1}U_i \to U_i \times \mathcal{R}^n$ such that:

(1) $\{U_i\}_{i \in \Lambda}$ is an indexed open cover of B;

(2) for each $i \in \Lambda$ the diagram

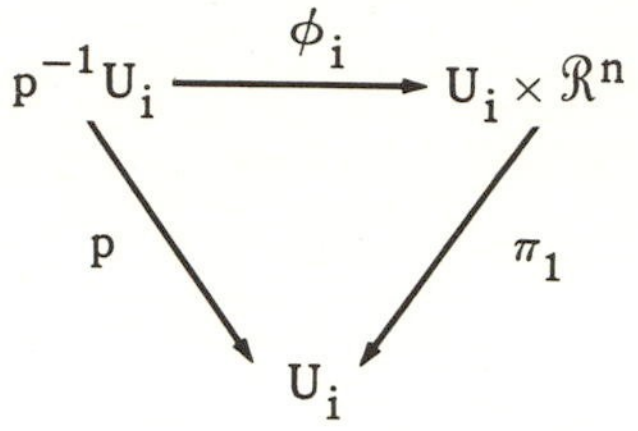

commutes, π_1 being the projection;

(3) for each $i, j \in \Lambda$ there are maps $g_{ji} : U_i \cap U_j \to GL(n)$ such that the homeomorphism $\phi_j \circ \phi_i^{-1}$ of $(U_i \cap U_j) \times \mathcal{R}^n$ is given by

$$(x, y) \mapsto (x, g_{ji}(x)\, y) \ .$$

The elements (ϕ_i, U_i) are called *local trivializations*; the g_{ij} are *transition functions*.

A maximal vector bundle atlas on (p, E, B) is a *vector bundle structure*. The quadruple (p, E, B, Φ) is a *vector bundle*.

(1.2) Vector bundles with "additional structure" of various kinds are defined by imposing restrictions on the family of transition functions. If the base space B is a polyhedron, for example, it makes sense to require that each map $g_{ji} : U_i \cap U_j \to GL(n)$ be PD, i.e., smooth $(= C^\infty)$ on each simplex of some triangulation of $U_i \cap U_j$. A *PD vector bundle atlas* is a vector bundle atlas having this property. A maximal PD vector bundle atlas Φ_0 is called a *PD vector bundle structure*; the quadruple (p, E, B, Φ_0) is a *PD vector bundle*.

(1.3) Let (p, E, B, Φ_0) be a PD vector bundle over a polyhedron B. It is easily proved that B has a triangulation T and Φ_0 has a subatlas Φ_1 with the following property: for every simplex Δ of T there exists a local trivialization $(\phi_i, U_i) \in \Phi_0$ such that $\Delta \subset U_i$, and if also $\Delta \subset U_j$, $(\phi_j, U_j) \in \Phi_0$, then the map

$$g_{ji} \mid \Delta : \Delta \to GL(n)$$

is smooth. Conversely, given a triangulation T of B and a vector bundle atlas Φ_1 on (p, E, B) having this property, there is a unique PD vector bundle atlas $\Phi_0 \supset \Phi_1$. Such a triangulation T is called a *base triangulation* of the PD vector bundle (P, E, B, Φ_0); the subatlas Φ_1 is *adapted* to T. In this case each map

$$\phi_i \circ \phi_j^{-1} : \Delta \times \mathcal{R}^n \to \Delta \times \mathcal{R}^n$$

is a diffeomorphism. Therefore $p^{-1}\Delta$ has a unique differentiable structure (with "corners") making each map $\phi_i : p^{-1}\Delta \to \Delta \times \mathcal{R}^n$ a diffeomorphism, for each simplex Δ of T and each $(\phi_i, U_i) \in \Phi_1$ with $\Delta \subset U_i$.

For $k = 0, 1$ let

$$x = (p_k, E_k, B_k, \Phi_k)$$

be a vector bundle. A *vector bundle map* $f : x_0 \to x_1$ is a map $f : E_0 \to E_1$ which covers a map $g : B_0 \to B_1$, such that for each $(\phi, U) \in \Phi_1$, the pair $(f^*\phi, g^{-1}U) \in \Phi_0$, where $f^*\phi : p_0^{-1}(g^{-1}U) \to (g^{-1}U) \times \mathcal{R}^n$, is defined by

$$y \mapsto (p_0(y), \pi_2\phi f(y)) ,$$

and $\pi_2 : U \times \mathcal{R}^n \to \mathcal{R}^n$ is the projection.

Now suppose that B_0 and B_1 are polyhedra. If $\Phi'_1 \subset \Phi_1$ is a PD (vector bundle) structure on x_1, and if $g : B_0 \to B_1$ is PL, then the set of local trivializations of the form

$$(f^*\phi, g^{-1}U), (\phi, U) \in \Phi'_1$$

is a PD atlas for x_0 and hence is contained in a unique PD structure $\Phi'_0 = f^*\Phi'_1$ on x_0. We call Φ'_0 the PD structure *induced* by f.

Suppose for $k = 0, 1$ that $x_k = (p_k, E_k, B_k, \Phi_k)$ is a PD vector bundle. A bundle map $f : \xi_0 \to \xi_1$ is a *PD bundle map* provided the induced map $g : B_0 \to B_1$ is PL, and also that $f^*\Phi_1 = \Phi_0$.

(1.4) THEOREM (Covering homotopy theorem for PD vector bundles). *Let* B *be a polyhedron,* $A \subset B$ *a subpolyhedron and* x *a* PD *vector bundle over* $B \times I$. *Let* $r : B \times I \to B \times 0$ *be the retraction* $r(x,t) = (x,0)$. *If* $f_0 : x|A \times I \to x|A \times 0$ *is a* PD *vector bundle retraction covering* $r|A \times I$, *then* f_0 *extends to a* PD *vector bundle retraction* $f : x \to x|B \times 0$ *covering* r.

Proof. The proof is almost identical to the usual simplex-by-simplex proof of the covering homotopy theorem for vector bundles over a simplicial complex, and is left to the reader.

The next result implies existence and uniqueness (up to PD vector bundle isomorphism) of PD structures on vector bundles.

(1.5) THEOREM. (a) *Let* B *be a polyhedron,* x *a vector bundle over* B, *and* Ψ_0 *a* PD *structure on* $x|A$. *Then there exists a* PD *structure* Ψ *on* x *which restricts to* Ψ_0 *over* A.

(b) *Let* x_0, x_1 *be* PD *vector bundles over polyhedra* B_0, B_1 *and* $f : x_0 \to x_1$ *a vector bundle map covering a* PL *map* $g : B_0 \to B_1$. *Let* $A \subset B_0$ *be a subpolyhedron such that the restriction* $f_A : x_0|A \to x_1$ *is a* PD *vector bundle map. Then there exists a homotopy* $f_t : x_0 \to x_1$ *of vector bundle maps such that for all* $0 \le t \le 1$:

(i) f_t covers g;

(ii) $f_{tA} = f_A$;

and

(iii) $f_1 : x_0 \to x_1$ *is a* PD *vector bundle map.*

Proof. We prove (b) first. By considering the commuting diagram

$$\begin{array}{ccc} x_0 & \longrightarrow & f^*x_1 \\ & \searrow \quad \swarrow & \\ & x_1 & \end{array}$$

of bundle maps, we may assume $B_0 = B_1 = B$, and $g = 1_B$. The standard induction on $\dim(B-A)$, and simplex-by-simplex proof of the inductive step, permit us to assume that B is a simplex Δ and $A = \partial\Delta$. The covering homotopy theorem implies that x_0 and x_1 are isomorphic to the trivial PD vector bundle $e = (\pi_1, \Delta\times\mathcal{R}^n, \Delta)$ (whose PD atlas contains the identity map of $\Delta\times\mathcal{R}^n$). Therefore we may assume $x_0 = x_1 = e$. Then the bundle map $f : e \to e$ is determined by a map $h : \Delta \to GL(n)$ which is PD on $\partial\Delta$. A homotopy $h_t : \Delta \to GL(n)$ of $h = h_0$ such that $h_t = h$ on $\partial\Delta$ and h_1 is PD determines the desired homotopy f_t. This proves (b).

To prove (a) we may assume $B = \Delta$, $A = \partial\Delta$, and x is a trivial vector bundle. Let $f : x \to e$ be a vector bundle isomorphism covering 1_Δ. By (b) the vector bundle map $f_{\partial\Delta} : x|\partial\Delta \to e|\partial\Delta$ is homotopic, through vector bundle maps, to a PD vector bundle map $g : (x|\partial\Delta, \Psi_0) \to e|\partial\Delta$. The homotopy from $f_{\partial\Delta}$ to g extends to a homotopy $f_t : x \to e$ of vector bundle maps covering 1_Δ. The PD structure Ψ on x induced by f_1 extends Ψ_0. The proof of 1.4 is complete.

(1.6) *Triangulations of* PD *vector bundles*

To each PD vector bundle $x = (p, E, B, \Phi)$ we associate a class of triangulations of the space E, as follows. A triangulation T of E is called *compatible* with x (or with Φ) if there exists a triangulation T_0 of B and an atlas $\Phi_0 \subset \Phi$ adapted to T (see 1.3) such that for each simplex $\Delta \subset B$ of T_0, $p^{-1}\Delta$ is a subcomplex of T, and in addition each simplex of $p^{-1}\Delta$ is smoothly embedded in the smooth manifold $p^{-1}\Delta$. In other words T induces a smooth triangulation of $p^{-1}\Delta$.

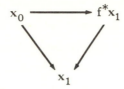

This means that if $(\phi, U) \epsilon \Phi_0, \Delta \subset U$, and σ is a simplex of T in $p^{-1}\Delta$, then $\phi|\sigma : \sigma \to \Delta \times \mathcal{R}^n$ is a smooth embedding.

Such a triangulation of E always exists. This is proved by applying J.H.C. Whitehead's theory of smooth triangulations to $\Delta \times \mathcal{R}^n$, extending a given smooth triangulation of $\partial\Delta \times \mathcal{R}^n$ which corresponds via a local trivialization to a compatible triangulation of $x|\partial\Delta$. Compatible triangulations of x are unique up to PL isomorphism.

If the full strength of Whitehead's theory is used, as developed for example in I, Theorem 13.4, then the compatible triangulation T of E can be chosen to have the following property. The map $p : E \to B$ is PL in the PL structure determined by T, and (p, E, B) is a locally trivial bundle in the PL category, with fibre the PL manifold $\mathcal{R}^n$. If T′ is another compatible triangulation of E having the same property, then there exists an isotopy $f_t : E \to E$ covering 1_B, having the following properties: $f_0 = 1_E$; $f_1 : (E, T') \to (E, T)$ is a PL isomorphism; and $f_t|p^{-1}\Delta$ is a PD isotopy from the PL structure induced by T′ into the differential structure induced by x, for each simplex Δ of T_0. Moreover $f_t|p^{-1}\Delta$ is the identity if T and T′ induce the same PL structure on $p^{-1}\Delta$. (For the existence of compatible triangulations use 13.4 of I, with $M = \Delta \times \mathcal{R}^n$ and $\phi = \pi_2 : \Delta \times \mathcal{R}^n \to \Delta$, etc. For isotopy either I, 13.4 or the covering homotopy theorem for PL bundle maps can be used.)

If B is a PL manifold, the PL structure on E induced by a compatible triangulation makes E a PL manifold.

(1.7) *Whitney sums*

Let $x = (p, E, M, \Phi)$ and $y = (q, F, N, \Psi)$ be PD vector bundles. The product PD vector bundle $x \times y = (p \times q, E \times F, M \times N, \Theta)$ has the PD vector bundle structure Θ determined by the PD vector bundle atlas

$$\{(\phi \times \psi, U \times V) | (\phi, U) \epsilon \Phi, (\psi, V) \epsilon \Psi\} .$$

It is easily verified that if E and F have PL structures compatible with Φ and Ψ, the product PL structures on $E \times F$ is compatible with Θ.

Suppose $M = N$ and K is the diagonal in $M \times M$, identified as usual with M. Then $x \times y | K$ is the *Whitney sum* $x \oplus y$, a PD vector bundle over M. If E and F have PL structures compatible with x and y, there is induced a natural PL structure on the total space $E(x \oplus y)$ of $x \oplus y$.

The total space of $x \oplus y$ is canonically identified with the total space of the PD vector bundle $y_0 = p^* y$ over E, induced from y by the projection $p : E \to M$ of x. It is easily verified that the PL structure just defined in $E(y_0) = E(x \oplus y)$ is compatible with $x \oplus y$.

Differentiable vector bundles

Let $x = (p, E, M, \Phi)$ be a vector bundle over a topological manifold M. Let α be a differential structure on M. A subatlas $\Phi_0 \subset \Phi$ is a *smooth atlas on* x *over* α provided each transition function of Φ_0,

$$g_{ji} : (U_i \cap U_j)_\alpha \to GL(n)$$

is a smooth map. A maximal smooth subatlas Φ_1 of Φ is a *differentiable bundle structure on* x *over* α. A *differentiable vector bundle* (x, α, Φ_1) consists of a vector bundle together with a differential structure α on B and a differentiable vector bundle structure Φ_1 on x over α.

There is a unique differential structure β on E making each local trivialization

$$\phi : (p^{-1} U)_\beta \to U_\alpha \times \mathcal{R}^n$$

of Φ_1 a diffeomorphism. The map $p : E_\beta \to M_\alpha$ is smooth of rank = dim M.

Let Φ_1 be a differentiable vector bundle atlas on $x = (p, E, M)$ over α, and β the corresponding differential structure on E. Let $g : M'_{\alpha'} \to M_\alpha$ be a smooth map and $x' = (p', E', M')$ the induced vector bundle. It is easily verified that the induced atlas $g^* \Phi_1$ is a differentiable vector bundle structure on x' over α'. Thus $g^* x$ has a canonical differentiable vector bundle structure induced from Φ_1.

Given differentiable vector bundles $(p', E', M', \alpha', \Phi')$ and (p, E, M, α, Φ), a vector bundle map

$$\begin{array}{ccc} E' & \xrightarrow{\ f\ } & E \\ {\scriptstyle p'}\downarrow & & \downarrow{\scriptstyle p} \\ M' & \xrightarrow{\ g\ } & M \end{array}$$

is a differentiable vector bundle map if the following two conditions are satisfied. First $g : M'_{\alpha'} \to M_\alpha$ is smooth. Second, for every $(\phi', U') \in \Phi'$ and $(\phi, U) \in \Phi$ such that $g(U') \subset U$, there is a smooth map $h : U' \to GL(n)$ such that the map

$$\phi \circ (\phi')^{-1} : U' \times \mathcal{R}^n \to U \times \mathcal{R}^n$$

is given by the formula

$$(x, y) \mapsto (g(x), h(x)y) .$$

If β' and β are the induced differential structures on E' and E, this is equivalent to saying that

$$f : E'_{\beta'} \to E_\beta \text{ is a smooth map.}$$

(1.8) THEOREM (Covering homotopy theorem for differentiable vector bundles). *Let* $x = (p, E, M_\alpha \times I, \Phi)$ *be a differentiable vector bundle where* M_α *is a smooth manifold. Let* $A \subset M$ *be a closed set and* $U \subset M$ *an open neighborhood of* A. *Let*

$$f_0 : x|U_\alpha \times I \to x|U_\alpha \times 0$$

be differentiable vector bundle retraction covering the retraction $r|U \times I$, *where* $r : M \times I \to M \times 0$ *is defined by* $r(x_1 t) = (x_1 0)$. *Then there exists a differentiable vector bundle retraction* $f : x \to x|M_\alpha \times 0$ *which covers* r *and extends* $f_0|V \times I$ *for some neighborhood* $V \subset M$ *of* A.

We leave the proof to the reader; for example, the proof of 1.9(b) below can be combined with the standard covering homotopy theorem for vector bundles.

The next result is the uniqueness and existence theorem for differentiable vector bundle structures.

(1.9) THEOREM. (a) *Let* M_α *be a smooth manifold and* $x = (p, E, M, \Phi)$ *a vector bundle. Let* $A \subset M$ *be a closed set,* $U \subset M$ *a neighborhood of* A, *and* Ψ_0 *a differentiable vector bundle structure on* $x|U$ *over* U_α. *Then there is a differentiable vector bundle structure* Ψ *on* x *over* M_α *such that* $\Psi|V = \Psi_0|V$ *for some neighborhood* V *of* A *in* U.

(b) *Let* $x_i = (p_i, E_i, M_i, \Phi_i)$ *be differentiable vector bundles over smooth manifolds* M_i *for* $i = 0, 1$. *Let* $f : x_0 \to x_1$ *be a vector bundle map covering a smooth map* $g : M_0 \to M_1$. *Suppose* $U \subset M_0$ *is a neighborhood of a closed set* $A \subset M_0$ *such that* $f_U : x_0|U \to x_1$ *is a differentiable vector bundle map. Then there exists a homotopy* $f_t : x_0 \to x_1$ *of vector bundle maps such that* $f_0 = f$, $f_t = f$ *over a neighborhood* V *of* A *in* U, *and* $f_1 : x_0 \to x_1$ *is a differentiable vector bundle map.*

Proof. A proof analogous to that of 1.5 is not difficult. Alternatively, to prove (b) for example, observe that vector bundle maps over g correspond to sections of a certain bundle y over M_0 with fibre GL(n), and group GL(n) acting by conjugation. y has a natural differentiable bundle structure (in the sense of Steenrod [28]). A differentiable section of y corresponds to a differentiable bundle map, and now (b) follows from standard approximation theorems. To prove (a), let $z = (q, E_0, G)$ be a differentiable vector bundle over a Grassmann manifold such that x is induced from z by a map $g : M \to G$, covered by a bundle map $f : x \to z$. We may assume g differentiable. By (b) we may assume that $f_V : x|V \to z$ is a differentiable vector bundle map for some neighborhood $V \subset U$ of A. Then the differentiable vector bundle structure Ψ induced by f from z satisfies (a) of 1.9.

(1.10) *Differentiable and* PD *vector bundles*

Note that if M is a PL manifold and α a smoothing of M, then a differentiable vector bundle (P, E, M, Φ) over α is also a PD vector bundle. (More precisely, Φ is contained in a unique PD vector bundle structure Φ_0.) We describe this situation by saying that the smooth vector bundle structure *induces* the PD vector bundle structure. If β is the differential structure on E induced by Φ and α, we shall also say that (β,α) is a *smoothing* of the PD vector bundle (P, E, M, Φ_0). Similarly, a differentiable vector bundle map is also a PD vector bundle map if the map of base spaces is simultaneously smooth and PL. Examples of such maps include: identity maps; projections $M_\alpha \times I \to M_\alpha$; inclusions $M_\alpha \times \{0\} \to M_\alpha \times \mathcal{R}^n$, etc.

(1.11) LEMMA. (a) *In Theorem* 1.9(a), *suppose* M *is a* PL *manifold and* α *a (compatible) smoothing of* M. *Suppose also that* $\Phi_1 \subset \Phi$ *is a* PD *vector bundle structure, and that* Φ_0 *induces* $\Phi_1|U$. *Then* Ψ *can be chosen so that* $\Psi \subset \Phi_1$.

(b) *In* 1.9(b), *suppose that* M_0 *and* M_1 *are* PL *manifolds with smoothings, and that* f *is a* PD *vector bundle map. Then the homotopy* f_t *can be chosen so that the map* $x_0 \times I \to x_1$, $(x, t) \mapsto f_t(x)$, *is a* PD *vector bundle map.*

Proof. The proof of 1.9 also proves 1.11.

Smoothings of total spaces

For the next lemmas, let $x_i = (p_i, E_i, M, \Phi_i)$ be a PD vector bundle over a PL manifold M, $i = 0, 1$. Let $f : x_0 \to x_1$ be an isomorphism of PD vector bundles covering 1_M. Suppose E_0 has a PL structure compatible with Φ_0.

(1.12) LEMMA. *Let* α *be a differential structure on* E_1. *The induced differential structure* $f^*\alpha$ *on* E_0 *is compatible with the* PL *structure on* E_0 *in each of the following cases:*

(a) E_1 *has a* PL *structure compatible with* α *and* $f: E_0 \to E_1$ *is* PL;

(b) α *is compatible with the* PD *vector bundle structure* Φ_1.

Proof. Left to the reader.

(1.13) LEMMA. *Let* α *be a smoothing of* M *and* β_i *a differential structure on* $E_i (i=0,1)$ *such that* (β_i, α) *is a smoothing of the* PD *vector bundle* x_i (see 1.10). *Let* $f: x_0 \to x_1$ *be a* PD *vector bundle isomorphism over* 1_M. *Then* $f^*\beta_1$ *and* β_0 *are concordant smoothings of* E_0.

Proof. Let $f_t: x_0 \to x_1$ be an isotopy of PD vector bundle maps as in 1.11, with $f_0 = f$ and f_1 an isomorphism of the differentiable vector bundles over M_α corresponding to β_0 and β_1. Let $F: E_0 \times I \to E_1 \times I$ be the map $F(x,t) = (f_t(x), t)$. Let ι be the standard smoothing of I; then 1.11(b) implies that $F^*(\beta_1 \times \iota)$ is compatible with the product PL structure on $E_0 \times I$. Since $f_1 : (E_0, \beta_0) \to (E_1, \beta_1)$ is a diffeomorphism, $F^*(\beta_1 \times \iota)$ is a concordance between β_0 and $f^*\beta_1$.

§2. *The map* $x^! : \mathcal{S}(M) \to \mathcal{S}(E)$

If M is a PL manifold then $\mathcal{S}^\bullet(M)$ denotes the set of smoothings of M, while $\mathcal{S}(M)$ denotes the set of concordance classes of smoothings. The concordance class of α is denoted by $[\alpha]$.

We recall from I, 7.8:

(2.1) THEOREM (Product smoothing theorem). *The natural map* $\mathcal{S}(M) \to \mathcal{S}(M \times \mathcal{R}^n)$ *is bijective for all* $n \in Z_+$.

Theorem 2.6 below generalizes this to arbitrary vector bundles over M. Other versions of this isomorphism are 2.6 and 2.7.

(2.2) LEMMA. *Let* E *and* M *be* PL *manifolds and* $x = (p, E, M, \Phi)$ *a* PD *vector bundle such that* Φ *is compatible with the* PL *structure on* E. *Let* α_0, α_1 *be smoothings of* M *and* β_0, β_1 *smoothings of* E *such that both* (β_0, α_0) *and* (β_1, α_1) *are smoothings of* x. *If* α_0 *and* α_1 *are concordant, then* β_0 *and* β_1 *are concordant.*

Proof. Consider the PD vector bundle $x \times I = (p \times 1, E \times I, M \times I, \Phi^*)$ induced from x by $\pi_1 : M \times I \to M$. Give $E \times I$ the product PL structure; this is compatible with Φ^*. Let $\alpha \in \mathcal{S}^\bullet(M \times I)$ be a concordance from α_0 on $M \times 0$ to α_1 on $M \times 1$. By 1.4(a) and 1.5(a) there exists a differentiable vector bundle structure Ψ on $x \times I$ over $(M \times I)$ which induces Φ^*, and such that if β is the smoothing of $E \times I$ corresponding to Ψ, then $(E \times i, \beta_i)$ is a smooth submanifold of $(E \times I, \beta)$ for $i = 0, 1$. Then β is a concordance from β_0 to β_1.

(2.3) Let $x = (p, E, M, \Phi)$ be a PD vector bundle over a PL manifold M. Suppose E has a PL structure compatible with Φ, making E a PL manifold.

For every smoothing α of M, there exists by 1.9(a) and 1.11(a) a smoothing β of E such that (β, α) is a smoothing of x. By 2.2 the concordance class of β is uniquely determined by that of α. The correspondence $[\alpha] \mapsto [\beta]$ defines a map denoted by

$$x^! : \mathcal{S}(M) \to \mathcal{S}(E) .$$

The truth of the following statement is obvious:

(2.4) LEMMA. *Let* $e = (p, M \times \mathcal{R}^n, M)$ *be a trivial* PD *vector bundle over a* PL *manifold* M. *Give* $M \times \mathcal{R}^n$ *the product* PL *structure. Let* ρ^n *be the natural smoothing of* $\mathcal{R}^n$. *Then the image of* $[\alpha]$ *under* $e^! : \mathcal{S}(M) \to \mathcal{S}(M \times \mathcal{R}^n)$ *is* $[\alpha \times \rho^n]$, *for all* $[\alpha] \in \mathcal{S}(M)$.

Now let x, y be PD vector bundles over M whose total spaces Ex, Ey have compatible PL structures. Let y_0 be the PD vector bundle over Ex induced from y by the projection p of x. We identify Ey_0 with $E(x \oplus y)$ as in 1.7. The same PL structure on Ey_0 is compatible with both y_0 and $x \oplus y$.

(2.5) LEMMA. *Let* M *be a* PL *manifold. Then the following diagram commutes*:

$$\begin{array}{ccc} \mathcal{S}(M) & \xrightarrow{x^!} & \mathcal{S}(Ex) \\ {\scriptstyle (x\oplus y)^!}\downarrow & & \downarrow{\scriptstyle y_0^!} \\ \mathcal{S}(E(x\oplus y)) & = & \mathcal{S}(Ey_0) \, . \end{array}$$

Proof. The diagram

$$\begin{array}{ccc} E(x\oplus y) & = & Ey_0 \\ \downarrow & & \downarrow \\ M & \longrightarrow & Ex \end{array}$$

commutes. Therefore if α, β, γ are smoothings of M, Ex and Ey_0 respectively, such that (β, α) is a smoothing of x and (γ, β) is a smoothing of y_0 (see 1.9) then (γ, α) is a smoothing of $x \oplus y$. This implies the lemma.

(2.6) THEOREM (Bundle smoothing theorem). *Let* $x = (p, E, M, \Phi)$ *be a* PD *vector bundle over a* PL *manifold* M. *Let* E *have a* PL *structure compatible with* Φ. *Then the map* $x^! : \mathcal{S}(M) \to \mathcal{S}(E)$ *is bijective.*

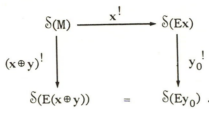

Proof. Let y be vector bundle over M such that $x \oplus y$ is isomorphic to the trivial vector bundle $e = (p, M \times \mathcal{R}^n, M)$. By (1.5) y has a PD structure and the corresponding PD structure on $x \oplus y$ is isomorphic to the canonical PD structure on e. Therefore we may assume y is a PD vector bundle, and let

$$f : x \oplus y \to e$$

be a PD vector bundle isomorphism over 1_M. Give $M \times \mathcal{R}^n$ the product PL structure. Then by 1.12 the map $f^* : \mathcal{S}(E(x \oplus y) \to \mathcal{S}(M \times \mathcal{R}^n)$ is defined, and by 1.13 the diagram

$$\begin{array}{ccc} \mathcal{S}(M \times \mathcal{R}^n) & \xleftarrow{\ f^*\ } & \mathcal{S}(E(x \oplus y)) \\ {\scriptstyle e^!}\uparrow & \nearrow {\scriptstyle (x \oplus y)^!} & \\ \mathcal{S}(M) & & \end{array}$$

commutes. By 2.5 the diagram

$$\begin{array}{ccc} \mathcal{S}(M) & \xrightarrow{\ (x \oplus y)^!\ } & \mathcal{S}(E(x \oplus y) = \mathcal{S}(Ey_0) \\ & {\scriptstyle x^!}\searrow & \uparrow {\scriptstyle y_0^!} \\ & & \mathcal{S}(Ex) \end{array}$$

commutes where $y_0 = p_x{}^* y$. Therefore $e^! : \mathcal{S}(M) \to \mathcal{S}(M \times \mathcal{R}^n)$ can be factored $e^! = f^* \circ y_0^! \circ x^!$. The product smoothing theorem states that $e^!$ is bijective. Therefore $x^!$ is injective, and $y_0^!$ is surjective since f^* is bijective. But x was arbitrary; hence $y_0^!$ is also injective, making $x^!$ bijective as well.

Germs of smoothings

Let V be a PL manifold and $X \subset V$ a subset. The set $\mathcal{S}^\bullet(V, X)$ of X-*germs of smoothings* of V is defined as the inverse limit of the maps $r_{ji} \mathcal{S}^\bullet(U_i) \to \mathcal{S}^\bullet(U_j)$ as U_i varies over open neighborhoods of X in V, and r_{ji} is induced by restriction if $U_j \subset U_i$. Two X-germs of smoothings are *concordant* if they have concordant representatives on some neighborhood of X. The set of concordance classes of X-germs is denoted by $\mathcal{S}(V, X)$.

Let $g : \mathcal{S}(V) \to \mathcal{S}(V, X)$ be the natural map.

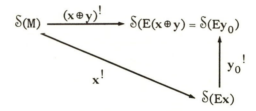

(2.7) LEMMA. *Let* $x = (p, E, M)$ *be a* PD *vector bundle over a* PL *manifold* M, *whose total space has a compatible* PL *structure. Identify* M *with the zero section of* E. *Then the map* $g : \mathcal{S}(E) \to \mathcal{S}(E, M)$ *is bijective.*

Proof. As discussed in 1.6, it may be assumed that the projection map $p : E \to M$ is PL locally trivial.

Let $A \subset M$ be a subpolyhedron and $W \subset E$ a neighborhood of $p^{-1}(A) \cup M$. There is a PL isotopy (not necessarily surjective) $f_t : E \to E$ such that:

$$f_0 = 1_E ,$$

$$f_1(E) \subset W ,$$

$$f_t(x) = x \text{ in a neighborhood of } p^{-1}(A) \cup M .$$

The existence of such an isotopy is easily proved for the case where M is a simplex Δ, $E = \Delta \times \mathcal{R}^n$, and $A = \partial\Delta$; the general case reduces to this by the PL local triviality of $p : E \to M$ and the usual simplex-by-simplex proof.

To prove g surjective let $W \subset E$ be a neighborhood of M and α a smoothing of W. Let f_t be an isotopy as above with $A = \phi$; then the M germ of $f_1^*\alpha$ is α.

To prove g injective, let α_0, α_1 be smoothings of E which are concordant on a neighborhood $V \subset E$ of M. We may choose a concordance $\beta' \epsilon \mathcal{S}^\bullet(V \times I)$ between $\alpha_0|V$ and $\alpha_1|V$ which extends to a smoothing β of a neighborhood $W \subset E \times I$ of $A = (M \times 0) \cup (M \times 1) \cup (V \times I)$. Let $f_t : E \times I \to E \times I$ be a PL isotopy such that $f_0 = 1$, $f_1(E \times I) \subset W$, and f_t is the identity on a neighborhood of A. Then $f_1^*\beta$ is a concordance between α_0 and α_1.

Note that we have actually proved that every M-germ of a smoothing of E extends to a smoothing of E; and the germ of a concordance between M-germs of smoothings of E extends to a concordance between the two smoothings of E.

Let $x = (p, E, M, \Phi)$ be as in 2.7. The composite map

$$\mathcal{S}(M) \xrightarrow{x^!} (E) \xrightarrow{g} \mathcal{S}(E, M)$$

will also be denoted by $x^!$.

(2.8) COROLLARY. *The map* $x^! : \mathcal{S}(M) \to \mathcal{S}(E, M)$ *is bijective.*

Proof. 2.7 and 2.6.

The map $x_X{}^! : \mathcal{S}(M, X) \to \mathcal{S}(E, X)$

Let $x = (p, E, M)$ be as in 2.7 and let $X \subset M$ be any subset. A map $x_X{}^! : \mathcal{S}(M, X) \to \mathcal{S}(E, X)$ is defined as follows. Let α be a smoothing of a neighborhood $W \subset M$ of X, let $[\alpha] \subset \mathcal{S}(M, X)$ be the concordance class of the X-germ of α. Let η be the restriction of x over W. Then $\eta^![\alpha] \in \mathcal{S}(p^{-1}W)$ is represented by a smoothing β of $p^{-1}W$. Since $p^{-1}W$ is a neighborhood of X in E, β represents an element $[\gamma] \in \mathcal{S}(E, X)$. Define $x_X{}^![\alpha] = [\gamma]$; it is easy to see that this definition is independent of the choices made. Since $\eta^! : \mathcal{S}(W) \to \mathcal{S}(p^{-1}W)$ is bijective no matter which neighborhood W is chosen, we have proved:

(2.9) LEMMA. *The map* $x_X{}^! : \mathcal{S}(M, X) \to \mathcal{S}(E, X)$ *is bijective.*

§3. *Linearization*

Let V be a PL manifold and $M \subset V$ a PL submanifold. A *linearization of* (V, M) is a PD vector bundle $x = (p, E, M)$ such that:

(1) E is a neighborhood in V of M,

(2) $p : E \to M$ is a retraction,

and

(3) the PL structure on E induced from V is compatible with x.

The set of linearizations of (V, M) is denoted by $\mathcal{L}^\bullet(V, M)$.

Two linearizations $x_i = (p_i, E_i, M)$ of (V, M), $i = 0, 1$, are *equivalent* if there exists a linearization $x = (p, E, M \times I)$ of $(V \times I, M \times I)$ such that for some neighborhood $W \subset V$ of M, $p(x,i) = (p_i(x), i)$ for $x \in (W \times i) \cap E$, $i = 0, 1$. The set of equivalence classes of linearizations of (V, M) is denoted by $\mathcal{L}(V, M)$.

Let $x = (P, E, M)$ be a PD vector bundle, and let $x \oplus e^q$ be the Whitney sum of x and the trivial PD vector bundle $M \times \mathcal{R}^q \to M$. We consider the total space of $x \oplus e^q$ to be $E \times \mathcal{R}^q$ and the projection to be the composition $p \circ \pi_1 : E \times \mathcal{R}^q \to E \to M$.

If x is a linearization of (V, M) then $x \oplus e^q$ is a linearization of $(V \times \mathcal{R}^q, M \times 0)$. In this way maps

$$\mathcal{L}(V, M) \to \mathcal{L}(V \times \mathcal{R}, M \times 0) \to \cdots \to \mathcal{L}(V \times \mathcal{R}^q, M \times 0) \to \cdots$$

are defined. The direct limit of this sequence of maps is the set of *stable linearizations* of (V, M), denoted by $\mathcal{L}_s(V, M)$. We denote by $s : \mathcal{L}(V,M) \to \mathcal{L}(V, M)$ the natural map.

The next theorem is stated in a stronger form than is needed. It would suffice to prove the following in place of (b): if V and M are smoothable then for some integer $q \geq 0$, $\mathcal{L}(V \times \mathcal{R}^q, M \times 0) \neq \emptyset$. The proof of this uses embedding theorems which are considerably simpler than those used in the proof of (b).

(3.1) LEMMA. *Let* (V, M) *be a* PL *manifold pair such that* V *is smoothable.*

(a) *If* $\mathcal{L}(V, M)$ *is not empty, then* M *is smoothable.*

(b) *If* M *is smoothable and* $\dim V > \frac{3}{2}(\dim M) + 1$, *then* $\mathcal{L}(V, M) \neq \emptyset$.

Proof. (a) Let $x = (p, E, M)$ be a linearization of (V, M). Since E is open in V, $\mathcal{S}(E) \neq \emptyset$. By 2.6, $\mathcal{S}(M) \neq \emptyset$.

(b) We may assume the inclusion $M \to V$ is a homotopy equivalence. Suppose V and M have smoothings β, α respectively. By Haefliger [4], the dimension restriction implies that the inclusion $M \to V$ is homotopic to a smooth embedding $f: M_\alpha \to V_\beta$. V_β has a smooth triangulation T making $f(M)$ a subcomplex of Γ, and such that $f: M \to V_T$ is PL. (See I, Section 13.) By uniqueness of smooth triangulations there is a PD isotopy $g_t: V_T \to V$ such that $g_0 = 1_V$, and $g_1: V_T \to V$ is a PL homeomorphism between the PL structure defined by T and the original PL structure on V. Then $h = g_1 \circ f: M \to V$ is a PL embedding homotopic to the inclusion such that $h(M)$ is a smooth submanifold of $(V, (g_1^{-1})^*\beta)$. Therefore we may suppose the inclusion $M \to V$ is homotopic to a smooth embedding $f: M_\alpha \to V_\beta$ such that $f: M \to V$ is a PL embedding. By Hudson [10] the dimension restriction implies that there is a PL isotopy $u_t: V \to V$ such that $u_0 = 1_V$ and $u_1|M = f$. Then $\gamma = u_1^*\beta$ is a smoothing of V in which M_α is a smooth submanifold. Now M_α has a differentiable normal vector bundle x in V_j and by the tubular neighborhood theorem we may consider x a linearization of (V, M). This proves (3.1).

The dimension restriction in 3.1(b) cannot be greatly relaxed. In [7] there is given an example of a locally flat PL submanifold $M^4 \subset S^7$ such that (S^7, M^4) has no linearization; but every four dimensional PL manifold can be smoothed (Cairns [2]).

(3.2) THEOREM. *Let* M *be a* PL *manifold,* $\partial M = \emptyset$. *Let* $M_\Delta \subset M \times M$ *denote the diagonal. Then the following statements are pairwise equivalent*:

(a) M *is smoothable*;

(b) $\mathcal{L}(M \times M, M_\Delta) \neq \emptyset$;

(c) $\mathcal{L}_s(M \times M, M_\Delta) \neq \emptyset$.

Proof. (a) $\Longrightarrow$ (b): If α is a smoothing of M then M_Δ is a smooth submanifold of $M_\alpha \times M_\alpha$.

(b) $\Longrightarrow$ (c): Obvious.

(c) $\Longrightarrow$ (a): Suppose $\mathcal{L}_s(M\times M, M_\Delta) \neq \emptyset$. Then for some $s > 0$, $\mathcal{L}(M\times M\times\mathcal{R}^s,\ \Delta\times 0) \neq \emptyset$. Put $V = M \times R^s$. Then $\mathcal{L}(V\times V, V_\Delta) \neq \emptyset$ because there is a PL homeomorphism

$$\phi : (V\times V, V_\Delta) \to ((M\times M\times R^s)\times R^s, (M_\Delta\times 0)\times R^s) .$$

By 2.1 it suffices to prove V smoothable. Therefore, replacing M by V, it suffices to prove:

$$\mathcal{L}(M\times M, M_\Delta) \neq \emptyset \Longrightarrow \mathcal{S}(M) \neq \emptyset .$$

Let $x = (p, E, M_\Delta)$ be a linearization of $(M\times M, M_\Delta)$, where $E \subset M\times M$ is a neighborhood of M_Δ and $p: E \to M_\Delta$ is a retraction. Stable uniqueness of PL normal bundles [8; 13; 16] implies that for $q \geq 0$ there is a PL embedding $g: E\times\mathcal{R}^q \to M\times M\times\mathcal{R}^q$ making the following diagram commute:

$$\begin{array}{ccc}
M_\Delta\times 0 & = & M_\Delta\times 0 \\
\downarrow & & \downarrow \\
E\times\mathcal{R}^q & \xrightarrow{\ g\ } & M\times M\times\mathcal{R}^q \\
\downarrow & & \downarrow \\
M_\Delta & = & M_\Delta
\end{array}$$

where $\lambda(x,y,z) = (x,x)$ and unlabeled maps are inclusions. (This is because the map $M\times M \to M_\Delta$, $(x,y) \mapsto (x,x)$, restricts to a PL bundle in some neighborhood of M_Δ.) Since $p\pi_1 : E\times\mathcal{R}^q \to M_\Delta$ is a PD vector bundle, the map g induces a linearization of $(M\times M\times\mathcal{R}^q, M_\Delta\times 0)$ whose projection is given by $(x,y,z) \mapsto (x,x,0)$. From a PL homeomorphism like ϕ above, we obtain a linearization of $((M\times\mathcal{R}^q)\times(M\times\mathcal{R}^q), (M\times\mathcal{R}^q)_\Delta)$ whose projection is: $(x,y) \mapsto (x,x)$. Since it suffices to prove $M\times\mathcal{R}^q$ smooth, it suffices to prove: M is smoothable provided there exists a linearization $x = (p, E, M)$ of $(M\times M, M_\Delta)$ such that $p(x,y) = (x,x)$.

We assume without loss of generality that $M \subset \mathcal{R}^s$, and there is PL retraction $r: N \to M$ of an open neighborhood $N \subset \mathcal{R}^s$ of M. Let

$$W = \{(x,v) \in M \times \mathcal{R}^s | x + v \in N, \text{ and } (r(x+v), x) \in E\} .$$

Let $E(r^*x)$ be the total space of the PD vector bundle over N induced by $r : N \to M$ from x, identifying M with M_Δ in the natural way. Give $E(r^*x)$ its natural PL structure. By definition

$$E(r^*x) = \{(y,z,w) \in N \times N \times M | (z,w) \in E, r(y) = z\} .$$

Define a map

$$f : W \to E(r^*x), f(x,v) = (x+v, r(x+v), x) .$$

Then f is injective and PL. Since the manifolds W and $E(r^*x)$ have the same dimension f maps W onto an open subset of $E(r^*x)$ by a PL homeomorphism.

The base space N of r^*x is smoothable, being open in $\mathcal{R}^s$. Hence $E(r^*x)$ is smoothable by 1.9(a). Therefore W is smoothable. Since $W \subset M \times \mathcal{R}^k$ is an open set containing $M \times 0$, it follows from 2.8 that M is smoothable. The proof of 3.2 is complete.

The map $\Phi : \mathcal{L}_s(V, M) \times \mathcal{S}(M) \to \mathcal{S}(V, M)$

Let V be a PL manifold and M a PL submanifold. A map

$$\Phi^\bullet : \mathcal{L}^\bullet(V, M) \times \mathcal{S}^\bullet(M) \to \mathcal{S}(V, M)$$

is defined by setting

$$\Phi^\bullet(x, \alpha) = x^![\alpha] ;$$

it is easy to see that $\Phi^\bullet(x,\alpha)$ depends only on the equivalence class of x and the concordance class of α. Therefore $\Phi^\bullet$ induces a map

$$\Phi : (V, M) \times \mathcal{S}(M) \to \mathcal{S}(V, M) .$$

It is clear that the following diagram commutes:

$$(3.3) \qquad \begin{array}{ccc} \mathcal{L}(V, M) \times \mathcal{S}(M) & \xrightarrow{\Phi} & \mathcal{S}(V, M) \\ \downarrow & & \downarrow \\ \mathcal{L}(V \times \mathcal{R}^q, M \times 0) \times \mathcal{S}(M \times 0) & \xrightarrow{\Phi} & \mathcal{S}(V \times \mathcal{R}^q, M \times 0) \end{array} ,$$

where the vertical maps are the natural ones. The map $\mathcal{S}(V, M) \to \mathcal{S}(V \times \mathcal{R}^q, r \times 0)$ is bijective by (2.9); hence Φ induces a map

$$\Phi : \mathcal{L}_s(V, M) \times \mathcal{S}(M) \to \mathcal{S}(V, M) .$$

In order to state the next theorem we define a map of sets

$$f : A \times B \to C$$

to be a *complete pairing* if it satisfies the following three conditions:

(1) if any two of the sets A, B, C are nonempty so is the third;

(2) f is *right complete*:

for all $a \in A$ the map

$$f_a : B \to C, \quad b \mapsto f(a, b)$$

is bijective;

(3) f is *left complete*:

for all $b \in B$ the map

$$f^b : A \to C, \quad a \mapsto f(a, b)$$

is bijective.

(3.4) THEOREM. *Let* V *be a* PL *manifold and* $M \subset V$ *a* PL *submanifold such that* $\partial M = M \cap \partial V$. *Then the map*

$$\Phi : \mathcal{L}_s(V, M) \times \mathcal{S}(M) \to \mathcal{S}(V, M)$$

is a complete pairing.

Proof. Condition (1) in the definition of complete pairing follows from 1.9(a), and 3.1, 2.1. Condition (2) follows immediately from 2.9. It remains to prove (3):

$$\Phi^{[\alpha]} : \mathcal{L}_s(V, M) \to \mathcal{S}(V, M) \text{ is bijective}$$

for every smoothing α of M. Referring to diagram 3.3, it suffices to show that for q sufficiently large, the map

$$\Phi : \mathcal{L}(V \times \mathcal{R}^q, M \times 0) \times \mathcal{S}(M \times 0) \to \mathcal{S}(V \times \mathcal{R}^q, M \times 0)$$

is a complete pairing. This follows from:

(3.5) LEMMA. *If* $\dim V \geq 2 \dim M + 5$, *then*

$$\Phi : \mathcal{L}(V, M) \times \mathcal{S}(M) \to \mathcal{S}(V, M)$$

is a complete pairing.

To prove this, let $[x] \in \mathcal{L}(V, M)$ and $[\alpha] \in \mathcal{S}(M)$. The fact that $\Phi_{[x]} = x^! : \mathcal{S}(M) \to \mathcal{S}(V, M)$ is bijective is proved in 2.8 (and needs no dimension restriction). It remains to prove that $\Phi^{[\alpha]} : \mathcal{L}(V, M) \to \mathcal{S}(V, M)$ is bijective.

To prove surjectivity let β be a smoothing of a neighborhood of M in V; we may replace V by this neighborhood and so let β be a smoothing of V. Since $\dim V > 2 \dim M$ there is a smooth embedding $f : M_\alpha \to V_\beta$. Considered as a PD embedding $M \to V_\beta$, f is PD isotopic to a PL embedding $g : M \to V$, by the theory of smooth triangulation. (For there is a smooth triangulation of V_β making $f(M_\alpha)$ a subcomplex; and there is a PD isotopy of this triangulation taking it to a PL isomorphism with the original PL structure in V.)

The PL embedding $g : M \to V$ is homotopic, and hence PL isotopic, to the inclusion, because $\dim V \geq 2 \dim M + 3$. By Hudson [9] this isotopy extends to a PL isotopy of V. (It is here that the assumption on ∂M and ∂V is used.) It follows that there is a PD isotopy $h_t : V \to V_\beta$ such that $h_0 = 1_V$, $h_1(M) = f(M_\alpha)$. Give (V, M) the linearization x induced by f from the smooth tubular neighborhood of $f(M_\alpha)$ in V_β. Then $x^![\alpha] = [\beta]$, which means that $\Phi^{[\alpha]}[x] = [\beta]$.

To prove injectivity of $\Phi^{[\alpha]}$ let x_0 and x_1 be linearizations of (V, M) such that $x_0{}^![\alpha] = x_1{}^![\alpha]$. This means that there is a smoothing β of $V \times I$ with $M_\alpha \times 0$ and $M_\alpha \times 1$ smooth submanifolds, and x_i is the tubular neighborhood of $M_\alpha \times i$ in $\partial(V\times I)_\beta$, $i = 0, 1$. Because of the dimension assumptions, there is a smooth embedding $F : M_\alpha \times I \to (V\times I)_\beta$ extending the inclusion of $M_\alpha \times \partial I$, and F is PD isotopic to the inclusion keeping $M \times \partial I$ fixed, by a PD isotopy $G_t : V \times I \to (V\times I)_\beta$ keeping $V \times \partial I$ fixed. Thus $G_1(M\times I) = F(M_\alpha \times I)$. A tubular neighborhood of $F(M_\alpha \times I)$, pulled back by G_1, provides a concordance between x_0 and x_1. This completes the proof of 3.4.

By using more sophisticated embedding and isotopy theorems of Haefliger [4] and Hudson [10], 3.5 can be improved to

(3.6) LEMMA. *Let* α *be a smoothing of* M. *The map* $\Phi^{[\alpha]} : \mathcal{L}(V, M) \to \mathcal{S}(V, M)$, *sending* $[x]$ *to* $x^![\alpha]$, *is surjective if* $\dim V \geq \frac{3}{2}(\dim M + 1)$; *and injective if* $\dim V > \frac{3}{2}(\dim M + 1)$.

The map $\mathrm{Exp} : \mathcal{S}(M) \to \mathcal{L}(M\times M, M_\Delta)$

Let M be a PL manifold without boundary. To every smoothing α of M we define a class of linearizations of $(M\times M, M_\Delta)$ as follows.

Let $T(M_\alpha)$ denote the total space of the tangent vector bundle of M_α. Pick a smooth Riemannian metric for α. There is a neighborhood $W \subset T(M_\alpha)$ of the zero section so small that for each $x \in M$, the exponential map

$$\exp_x : W \cap T_x(M_\alpha) \to M_\alpha$$

is a diffeomorphism onto a convex neighborhood of x.

Let $r : T(M_\alpha) \to W$ be a smooth fibre preserving embedding which is the identity on a neighborhood of the zero section. Define a smooth embedding

$$e : T(M_\alpha) \to M \times M$$

by

$$e(Y) = (x, \exp r(Y)), \quad (Y \in T_x(M_\alpha)) .$$

Thus e maps the zero section onto the diagonal in the obvious way.

By means of e we obtain a linearization (q, E, M_Δ) of $(M\times M, M_\Delta)$: put $E = e(T(M_\alpha))$ and $q = d\circ p\circ e^{-1}$ where $d : M \to M\times M$ is the diagonal embedding and $p : T(M_\alpha) \to M$ is the bundle projection. Give (q, E, M_Δ) the unique smooth vector bundle structure making e^{-1} a smooth vector bundle isomorphism. In this way a map $\mathrm{Exp}_1^\bullet : \mathcal{S}^\bullet(M) \to \mathcal{L}^\bullet(M\times M, M)$ is defined.

The linearization $\mathrm{Exp}_1^\bullet(\alpha)$ of $(M\times M, M_\Delta)$ depends on the choice of the Riemannian metric, the neighborhood W, and the embedding r. Different choices, however, lead to an equivalent linearization, since any two smooth Riemannian metrics on M_α can be connected by an interval in the convex cone of such metrics, leading to an isotopy between their exponential maps on some neighborhood of the zero section; and any two embeddings such as r are smoothly isotopic rel a neighborhood of the zero section. Moreover, if α_0 and α_1 are isotopic smoothings of M, the resulting linearizations are equivalent. Therefore we have defined a map

$$\mathrm{Exp}_1 : \mathcal{S}(M) \to \mathcal{L}(M\times M, M_\Delta) .$$

Let $\Lambda : M\times M \to M\times M$ be the "flip" $(x,y) \mapsto (y,x)$. If we replace the map $e : T(M_\alpha) \to M\times M$ by $\Lambda\circ e : T(M_\alpha) \to M\times M$ in the definition of Exp_1, maps $\mathrm{Exp}_2^\bullet : \mathcal{S}^\bullet(M) \to \mathcal{L}^\bullet(M\times M, M_\Delta)$ and $\mathrm{Exp}_2 : \mathcal{S}(M) \to \mathcal{L}(M\times M, M_\Delta)$ are obtained.

(3.7) LEMMA. $\mathrm{Exp}_2 = \mathrm{Exp}_1$.

Proof. This is just the tubular neighborhood theorem applied to the "vertical" and "horizontal" tubular neighborhoods of M_Δ in $M_\alpha \times M_\alpha$, which define $\mathrm{Exp}_1^{[\alpha]}$ and $\mathrm{Exp}_2^{[\alpha]}$ respectively.

We denote by Exp the map

$$\mathrm{Exp} = \mathrm{Exp}_1 = \mathrm{Exp}_2 : \mathcal{S}(M) \to \mathcal{L}(M\times M, M_\Delta) .$$

Let α and ω be smoothings of M. Then $\omega \times \alpha$ is a smoothing of $M \times M$; we denote by $[\omega\times\alpha]_\Delta$ the concordance class of the M_Δ-germ of $\omega \times \alpha$. Thus $[\omega\times\alpha]_\Delta \in \mathcal{S}(M\times M, M_\Delta)$.

(3.8) LEMMA.

(a) $(\mathrm{Exp}[\alpha])^![\omega] = [\omega\times\alpha]$

(b) $[\omega\times\alpha]_\Delta = [\alpha\times\omega]_\Delta$.

Proof. (a) By definition, $(\mathrm{Exp}[\alpha])^![\omega]$ is the concordance class of the M_Δ-germ of a smoothing of the total space E of $(\mathrm{Exp}_1\alpha)^![\omega]$ which makes the vector bundle $x_0 = (p, E, M_\Delta)$ into a smooth vector bundle over M_ω. Such a smoothing can be obtained as follows. Let $g_t : M \to M_\alpha$ be a PD homotopy such that $g_0 = 1_M$, $g_1 : M_\omega \to M_\alpha$ is a smooth map, and $(x, g_t(x)) \in E$ for all $x \in M$, $t \in I$. Define $G_t : M_\Delta \to E \subset M \times M$ by $G_t(x,x) = (x, g_t(x))$. Extend G_t to a neighborhood N of M_Δ in E as follows. If $(x,y) \in E$ then (x,y) belongs to the fibre over (x,x) in the vector bundle $\exp_1(\alpha) = (q, E, M_\Delta)$. Hence we may think of (x,y) as a vector; (x,x) is the origin. Then define $G_t(x,y) = G_t(x,x) + (x,y)$, vector addition. More precisely, $G_t(x,y) = e(e^{-1}G_t(x,x) + e^{-1}(x,y))$, where $e : T(M_\alpha) \to M \times M$ defines the vector bundle structure on E. The homotopy G_t is well defined on a neighborhood N of M_Δ in E, and is in fact a PD isotopy $N \to M_\omega \times M_\alpha$. It is clear that the map e and $G_1 \circ e$ induce equivalent linearizations x_0, x_1 respectively of (N, M_Δ), hence $x^![\omega] = x_1^![\omega]$. That induced by $G_1 \circ e$ is a tubular neighborhood of the *smooth* submanifold $G_1(M) \subset M_\omega \times M_\alpha$, since $G_1(M)$ is the graph of the smooth map $g_1 : M_\omega \to M_\alpha$. This means that $x_1^![\omega] = (G_1^*[\omega\times\alpha])_\Delta$. Since G_1 is PD isotopic to the identity map of N, $(G_1^*[\omega\times\alpha])_\Delta = [\omega\times\alpha]_\Delta$, which proves (a).

To prove (b), we have from (a) and (3.7):

$$\begin{aligned}[\omega\times\alpha]_\Delta &= (\mathrm{Exp}_1[\alpha])^![\omega] = (\mathrm{Exp}_2[\alpha])^![\omega] = \Lambda^*((\mathrm{Exp}_1[\alpha])^![\omega]) \\ &= [\Lambda^*(\omega\times\alpha)]_\Delta = [\alpha\times\omega]_\Delta\ .\end{aligned}$$

Denote by $\mathbf{Exp} : \mathcal{S}(M) \to \mathcal{L}_s(M \times M, M_\Delta)$ the composition

$$\mathcal{S}(M) \xrightarrow{\ \mathrm{Exp}\ } \mathcal{L}(M \times M, M_\Delta) \xrightarrow{\ s\ } \mathcal{L}_s(M \times M, M_\Delta) .$$

(3.9) LEMMA. *The maps*

$$\Phi : \mathcal{L}_s(M \times M, M_\Delta) \times \mathcal{S}(M) \to \mathcal{S}(M \times M, M_\Delta)$$

and

$$\mathbf{Exp} : \mathcal{S}(M) \to \mathcal{L}_s(M \times M, M_\Delta)$$

satisfy the relation

$$\Phi(\mathbf{Exp}\ \alpha, \omega) = \Phi(\mathbf{Exp}\ \omega, \alpha)$$

for all $\alpha, \omega \in \mathcal{S}(M)$.

Proof. Note that α, ω denote *concordance classes* of smoothings of M.

It suffices to prove that

$$\Phi(\mathrm{Exp}\ \alpha, \omega) = \Phi(\mathrm{Exp}\ \omega, \alpha) .$$

By definition of Φ, this is equivalent to

$$(\mathrm{Exp}\ \alpha)^!\omega = (\mathrm{Exp}\ \omega)^!\alpha ,$$

which follows from 3.8.

(3.10) THEOREM. *Let* M *be a* PL *manifold without boundary. Then the map* $\mathbf{Exp} : \mathcal{S}(M) \to \mathcal{L}_s(M \times M, M_\Delta)$ *is bijective.*

Proof. This is a purely formal consequence of 3.4, 3.9, and the following fact.

(3.11) LEMMA. *Let* A, B, Z *be sets and* $\Phi : A \times B \to Z$, $e : B \to A$ *maps. Suppose that* Φ *is a complete pairing* (see 3.4), *and that*

$$\Phi(e(\alpha), \omega) = \Phi(e(\omega), \alpha)$$

for all $\alpha, \omega \in B$. *Then* e *is bijective.*

Proof. Fix $\alpha \in B$. The map $\omega \mapsto \Phi(e(\alpha),\omega)$ is a bijection $B \to Z$, since Φ is right complete. Therefore the map $\omega \mapsto \Phi(e(\omega),\alpha)$ is a bijection $B \to Z$. Because Φ is left complete, e must be bijective.

§4. *Classifications*

Theorem 3.10 says that the classification of smoothings of M is the same as the classification of stable linearizations of $(M \times M, M_\Delta)$. This last classification is translated to a homotopy problem via the theory of *block bundles* [23]. We do not repeat the definition of block bundle, but merely note the following facts.

If M is a locally flat PL submanifold of V, then a regular neighborhood of M in V determines an isomorphism class of block bundles over M; this class is independent of the choice of regular neighborhood. The functor which to each polyhedron P assigns the set of isomorphism classes of block bundles over P of fibre dimension n, is a homotopy functor and has a classifying space (or semi-simplicial complex) $B\,\widetilde{PL}_n$. (There is a semisimplicial group $\widetilde{PL}_n$ and $B\,\widetilde{PL}_n$ may be taken to be its classifying space.)

The operation of triangulating vector PD bundles induces a map $t_n : B\,O(n) \to B\,\widetilde{PL}(n)$, which we may take to be a fibration with the following property. Let

$$f : M \to B\,\widetilde{PL}(n)$$

be the classifying map of the normal block bundle of M in V. Then *equivalence classes of linearizations of* (M, V) *correspond bijectively to homotopy classes of liftings* $g : M \to B\,O(n)$ *of* f *over* t_n.

There are natural inclusions $B\,\widetilde{PL}(n) \to B\,\widetilde{PL}(n+1)$. Passing to the direct limit, we denote by $B\,\widetilde{PL}$ the classifying space for stable equivalence classes of block bundles. Equivalence classes of stable linearizations of (V, M) correspond bijectively to homotopy classes of liftings $g : M \to BO$ of $f_s : M \to B\,\widetilde{PL}$ over $t : B\,\widetilde{PL} \to BO$, where f_s is the composition $M \xrightarrow{f} B\,\widetilde{PL}_n \to B\,\widetilde{PL}$, and t is the direct limit of $\{t_n\}$.

The fibre of $t_n : BO(n) \to B\widetilde{PL}_n$ is denoted by $\widetilde{PL}_n/O_n$, and the fibre of t by $\widetilde{PL}/O$. (Later we shall put $\widetilde{PL}/O = \Gamma$.)

For a PL manifold M without boundary, let $f_M : M \to B\widetilde{PL}$ classify the stable normal block bundle of M_Δ in $M \times M$. Let $\mathcal{E}(M) \to M$ be the fibration induced by f_M from $t : BO \to B\widetilde{PL}$. Then 3.10 has the following interpretation.

(4.1) THEOREM. *Concordance classes of smoothings of* M *correspond bijectively to homotopy classes of sections of the bundle* $\mathcal{E}(M) \to M$ *whose fibre is* $\widetilde{PL}/O$. *The correspondence is natural for* PL *homeomorphisms.*

The last sentence means that if $g : N \to M$ is a PL homeomorphism, we may take $f_N = f_M \circ g$, and $\mathcal{E}(N) = g^*\mathcal{E}(M)$. Therefore there is a natural correspondence $g^\#$ between sections of $\mathcal{E}(M)$ and $\mathcal{E}(N)$, preserving homotopies. If a section σ of $\mathcal{E}(M)$ corresponds to a smoothing α of M, then $g^\#\sigma$ corresponds to $g^*\alpha$.

Another classification is obtained as follows. Let M be a PL submanifold of V; *assume* V *is smoothable and let* ω *be a fixed smoothing of* V. By 3.4, for every smoothing α of M there is a unique equivalence class of stable linearizations x of (V, M) such that $x^![\alpha] = [\omega]$. This correspondence $\alpha \mapsto x$ defines a bijection $\mathcal{S}(M) \to \mathcal{L}_s(V, M)$, which depends on $[\omega]$.

Consider the special case $V = M$. Then $\mathcal{L}_s(M, M)$ is the set of equivalence classes of stable linearizations of the trivial stable block bundle over M, and is canonically identified with homotopy classes of liftings of the constant map $M \to B\widetilde{PL}$ over $t : BO \to B\widetilde{PL}$. In other words

$$\mathcal{S}(M) \approx \mathcal{L}_s(M, M) = [M, \widetilde{PL}/O] ,$$

homotopy classes of maps $M \to \widetilde{PL}/O$. This proves the following classification theorem.

(4.2) THEOREM. *Let* M *be a smoothable* PL *manifold. For each concordance class* $\omega \in \mathcal{S}(M)$ *there is a bijection*

$$\Sigma_\omega : \mathcal{S}(M) \to [M, \widetilde{PL/O}] ,$$

with $\Sigma_\omega(\omega)$ *being the homotopy class of the constant map. If* $g: M_\omega \to N_\theta$ *is both a diffeomorphism and a* PL *homeomorphism, then the following diagram commutes*:

$$\begin{array}{ccc} \mathcal{S}(N) & \xrightarrow{\Sigma_\theta} & [N, \widetilde{PL/O}] \\ \downarrow g^* & & \downarrow g^\# \\ \mathcal{S}(M) & \xrightarrow{\Sigma_\omega} & [M, \widetilde{PL/O}] \end{array} .$$

From the point of view of smooth manifolds, 4.2 is unsatisfactory in that it refers to a specific compatible PL structure on the smooth manifold M_ω. In order to deal with *smooth* manifolds rather than *smoothed* PL manifolds we make the following definitions.

Let M be a topological manifold and ω a smoothing of M. A *resmoothing of* M_ω is a smoothing of M which has a smooth triangulation in common with ω. If α_0 and α_1 are resmoothings of M_ω, they are called *ω-concordant* if there is a smoothing α of $M \times I$ which induces α_i on $M \times i$ $(i=0,1)$, and which has a smooth triangulation $t: P \times I \to (M \times I)_\alpha$ where P is a PL manifold and $P \times I$ has the product PL structure. It is not assumed that t is level-preserving. It is clear that ω-concordance is a symmetric and reflexive relation on the set of differential structures on M; transitivity is easily proved using the theory of smooth triangulations. The set of ω-concordance classes of resmoothings of M_ω is denoted by $\mathcal{S}(M_\omega)$.

Let K be a PL structure on M compatible with ω; denote by M_K the corresponding PL manifold. A map $t_K : \mathcal{S}(M_K) \to \mathcal{S}(M_\omega)$ is defined by $t_K[\alpha] = [\alpha]_\omega$ = the ω-concordance class of the smoothing α.

Let L be another PL structure on M compatible with ω. From the theory of smooth triangulations we obtain a PL isomorphism $f: M_K \to M_L$

which is PD isotopic to 1_M as a PD map $M_K \to M_\omega$. There is induced a bijective map $q^*_{KL} : \mathcal{S}(M_L) \to \mathcal{S}(M_K)$ which does not depend on the choice of f.

The proof of the following lemma is left to the reader.

(4.3) LEMMA. (a) *The diagram*

$$\begin{array}{ccc} \mathcal{S}(M_L) & \xrightarrow{q^*_{KL}} & \mathcal{S}(M_K) \\ & {\scriptstyle t_L} \searrow \quad \swarrow {\scriptstyle t_K} & \\ & \mathcal{S}(M_\omega) & \end{array}$$

commutes.

(b) t_K *and* t_L *are bijective; hence* q^*_{KL} *is bijective.*

Let $\mathbf{D}$ be the category of smooth manifolds and diffeomorphisms, and $\mathbf{S}$ the category of pointed sets and maps. Define a contravariant functor $\mathcal{S} : \mathbf{D} \to \mathbf{S}$ by assigning to M_ω the set $\mathcal{S}(M_\omega)$ with base point $[\omega]_\omega$; and to a diffeomorphism $g : M_\omega \to N_\theta$, assign the map $\mathcal{S}(g) : \mathcal{S}(N_\theta) \to \mathcal{S}(M_\omega)$, $[\alpha]_\theta \to [g^*\alpha]_\omega$.

Let $\mathcal{T} : \mathbf{D} \to \mathbf{T}$ be the natural functor from $\mathbf{D}$ to the category $\mathbf{T}$ of topological spaces and continuous maps.

Let $\Gamma = \widetilde{PL}/O$ be the fibre of $BO \to B\widetilde{PL}$; let it also denote the functor $\Gamma : \mathbf{T} \to \mathbf{S}$ which assigns to a continuous map $h : X \to Y$ the map $h^\# : [Y, \Gamma] \to [X, \Gamma]$ of homotopy sets.

(4.4) THEOREM. *The functors* $\mathcal{S}$ *and* $\Gamma \circ \mathcal{T} : \mathbf{D} \to \mathbf{S}$ *are naturally equivalent. In other words, the homotopy functor* $[M, \Gamma]$, *pulled back to* $\mathbf{D}$, *is naturally equivalent to* $\mathcal{S}$.

Proof. For each smooth manifold M_ω, let $K = K_\omega$ be a compatible PL structure on M. Define a map

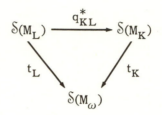

$$H = H(M_\omega) : \mathcal{S}(M_\omega) \to [M, \Gamma]$$

to be the composition

$$H : \mathcal{S}(M_\omega) \xrightarrow{t_K^{-1}} \mathcal{S}(M_K) \xrightarrow{\Sigma_\omega} [M, \Gamma] .$$

By 4.2 and 4.3 H is well defined independent of the choice of K, and is bijective. The naturality of H also follows easily from 4.2 and 4.3. Thus H is the required natural equivalence.

Let M, N be PL manifolds and ω, ξ respective smoothings of M, N. According to 4.4, if $f: M \to N$ is a continuous map, there is defined a map $\mathcal{S}(f): \mathcal{S}(N_\xi) \to \mathcal{S}(M_\omega)$. We proceed to explain $\mathcal{S}(f)$ directly. In doing so we shall identify $\mathcal{S}(M_\omega)$ with $\mathcal{S}(M)$ via a map t_K as in 4.3.

First consider the zero section $j: N \to N \times \mathcal{R}^q$. It is not hard to see that the map

$$\mathcal{S}(j) : \mathcal{S}(N_\xi) \to \mathcal{S}(N \times \mathcal{R}^q, \xi \times \rho^q)$$

is the map $[\beta] \mapsto [\beta \times \rho^q]$, where ρ^q is the natural smoothing on $\mathcal{R}^q$. By naturality, for $[\beta] \in \mathcal{S}(N_\xi)$,

$$\mathcal{S}(f)[\beta] = \mathcal{S}(j \circ f)[\beta \times \rho^q] .$$

Moreover, since $\mathcal{S}$ is known to be a homotopy functor, we may replace $j \circ f$ by any homotopic map. In particular, if q is sufficiently large we may find a smooth embedding $g: M_\alpha \to N_\xi \times \mathcal{R}^q$, homotopic to $j \circ f$; then

$$\mathcal{S}(f)[\beta] = \mathcal{S}(g)[\beta \times \rho^q] .$$

Now $g(M_\alpha)$ has a smooth normal bundle in $N_\xi \times \mathcal{R}^q$, which defines a linearization of $(N \times \mathcal{R}^q, g(M))$ (after a smooth triangulation of $N_\xi \times \mathcal{R}^q$ with $g(M)$ a subcomplex). Denote this linearization by $\mathfrak{n} = (p, E, g(M))$, where $E \subset N \times \mathcal{R}^q$ is an open set containing $g(M)$. Let β_0 be the restriction of $\beta \times \rho^n$ to E. By 2.8 there is a unique concordance class of smoothings γ of $g(M)$ such that $\mathfrak{n}^![\gamma] = [\beta_0]$.

(4.5) THEOREM. $\mathcal{S}(f)[\beta] = [g^*\gamma]$.

Proof. Left to the reader.

As another exercise the reader is invited to prove:

(4.6) THEOREM. *Let* $x = (p, E, M)$ *be a* PD *vector bundle, where* E *and* M *are* PL *manifolds whose* PL *structures are compatible with* x. *Let* $[\alpha] \in \mathcal{S}(M)$; *put* $[\theta] = x^![\alpha] \in \mathcal{S}(E)$. *Then the map*

$$\mathcal{S}(p) : \mathcal{S}(M_\alpha) \to \mathcal{S}(E_\theta)$$

is given by $[\beta] \mapsto x^![\beta]$.

As a corollary of 4.4 we have:

(4.7) THEOREM. *The functor* $\mathcal{S} : \mathbf{D} \to \mathbf{S}$ *from the category of smooth manifolds and diffeomorphisms, extends to a homotopy functor on* $\mathbf{D}$.

Remark. Let PL_n denote the semisimplicial group of PL homeomorphisms of $(R^n, 0)$; set $PL = \lim_{n\to\infty} PL_n$. There is a natural homotopy equivalence $B\,PL \to B\,\widetilde{PL}$. Therefore Γ can be identified with the fibre of $BO \to B\,PL$, sometimes denoted by PL/O. See also Section 6.

The abelian group structure on $\mathcal{S}(M_\omega)$

Let M be a PL manifold and ω a fixed smoothing of M. Let $d : M \to M \times M$ be the diagonal embedding. We define a binary operation $P_\omega : \mathcal{S}(M_\omega) \times \mathcal{S}(M_\omega) \to \mathcal{S}(M_\omega)$ by $P_\omega([\alpha], [\beta]) = \mathcal{S}(d)[\alpha \times \beta]$ where $\mathcal{S}$ is the functor of 4.7. Explicitly, $P_\omega([\alpha], [\beta])$ is represented by a smoothing γ of M such that

$$(\operatorname{Exp} \omega)^! [\gamma] = [\alpha \times \beta]_\Delta ,$$

the M_Δ germ of $\alpha \times \beta$. Equivalently, by 3.8, γ is determined by the equation

$$[\omega \times \gamma]_\Delta = [\alpha \times \beta]_\Delta \ .$$

(4.8) LEMMA. P_ω *is commutative and is a complete pairing.*

Proof. Follows from 3.8 and 3.10.

(4.9) THEOREM. *The operation* P_ω *makes* $\mathcal{S}(M_\omega)$ *into an abelian group whose identity element is* $[\omega]$. *If* $f : M_\omega \to N_\theta$ *is a continuous map, then*

$$\mathcal{S}(f) : \mathcal{S}(N_\theta) \to \mathcal{S}(M_\omega)$$

is a homomorphism of groups.

Proof. We have already proved commutativity, and clearly $[\omega]$ is a two-sided identity. Inverses exist since the pairing is complete. It remains to prove associativity.

Let $M_0 \subset M \times M \times M = M^{(3)}$ be the diagonal:

$$M_0 = \{(x, y, z) \in M \times M \times M \mid x = y = z\} \ .$$

Let x be a linearization of $(M^{(3)}, M_0)$ corresponding to a smooth tubular neighborhood of $(M_0)_\omega$ in $M_\omega \times M_\omega \times M_\omega$.

Let α, β, γ be concordance classes of smoothings of M. Denote $P_\omega(\alpha, \beta)$ by $\alpha\beta$, etc. It is not hard to verify that both $(\alpha\beta)\gamma$ and $\alpha(\beta\gamma)$ are equal to the concordance class δ determined by the condition that $x^!\delta$ is represented by the M_0 germ of the smoothing $\alpha \times \beta \times \delta$. Since $x^!$ is injective by 2.8, associativity is proved.

We must prove that if $f : M_\omega \to N_\theta$, then $\mathcal{S}(f)(\alpha\beta) = (\mathcal{S}(f)\alpha)(\mathcal{S}(f)\beta)$. It is sufficient to observe that the diagram

$$\begin{array}{ccc} M \times M & \xrightarrow{f \times f} & N \times N \\ \uparrow{\scriptstyle d_M} & & \uparrow{\scriptstyle d_N} \\ M & \xrightarrow{f} & N \end{array}$$

commutes, where d_M and d_N are the diagonal maps, and $\mathcal{S}(d_N f)(\alpha \times \beta) = \mathcal{S}(f)(\alpha\beta)$ while $\mathcal{S}((f \times f) d_M)(\alpha \times \beta) = (\mathcal{S}(f)\alpha)(\mathcal{S}(f)\beta)$.

We can now improve the homotopy functor $\mathcal{S} : \mathbf{D} \to \mathbf{S}$. Let $\mathbf{A}$ be the category of abelian groups and homomorphisms and $\mathbf{A} \to \mathbf{S}$ the forgetful functor.

(4.10) THEOREM. *The homotopy functor* $\mathcal{S} : \mathbf{D} \to \mathbf{S}$ *lifts to a homotopy functor* $\mathbf{D} \to \mathbf{A}$.

Proof. Follows from 4.9.

We shall denote by $\mathcal{S}$ the functor $\mathbf{D} \to \mathbf{A}$ determined by the group structure on $\mathcal{S}(M_\omega)$ specified by 4.9.

We emphasize that the *group structure on* $\mathcal{S}(M_\omega)$ *depends on the concordance class* $\omega \in \mathcal{S}(M)$. Any two such structures are isomorphic, however.

Abelian group structure on $[X, \Gamma]$

Recall that Γ denotes the fibre of the map $BO \to B\widetilde{PL}$ (also denoted by $\widetilde{PL}/O$). Let $\mathbf{P}$ denote the category of spaces having the homotopy type of polyhedra, and homotopy classes of maps. An object X of $\mathbf{P}$ is a topological space homotopy equivalent to a subcomplex of a rectilinear triangulation of a Euclidean space; equivalently, X is a homotopy equivalent to a finite dimensional, locally finite CW complex having countably many components.

Let $\Gamma : \mathbf{P} \to \mathbf{S}$ be the functor $X \mapsto [X, \Gamma]$.

(4.11) THEOREM. *The functor* $\Gamma : \mathbf{P} \to \mathbf{S}$ *lifts to a functor* $\mathbf{P} \to \mathbf{A}$.

Proof. Let P be an object of $\mathbf{P}$. Choose a homotopy equivalence $\phi : P \to Q$ from P to a polyhedron $Q \subset \mathcal{R}^n$. Let $N \subset \mathcal{R}^n$ be an open set admitting Q as a deformation retract. Let ν be the smoothing of N

induced by the inclusion $N \to \mathcal{R}^n$. The composition $P \xrightarrow{\phi} Q \subset N$ is a homotopy equivalence and defines a bijection

$$[P,\Gamma] \approx [N,\Gamma] .$$

From 4.4 there is a bijection

$$[N,\Gamma] \approx \mathcal{S}(N_\omega) .$$

By 4.10, $\mathcal{S}(N_\omega)$ is an abelian group; we denote it by $A(P)$.

Let P' be another object of **P** and let $\phi' : P' \to Q'$ be a homotopy equivalence to a polyhedron $Q' \subset \mathcal{R}^{n'}$; let $N' \subset \mathcal{R}^{n'}$ be an open set having $\phi'(P')$ as a deformation retract; and let ν' be the induced smoothing of N'. Given a map $g : P \to P'$, there is a unique homotopy class of maps $G : N \to N'$ such that the diagram

$$\begin{array}{ccc} P & \xrightarrow{g} & P' \\ {\scriptstyle\phi}\downarrow & & \downarrow{\scriptstyle\phi'} \\ N & \xrightarrow{G} & N' \end{array}$$

commutes up to homotopy. Define $A(g) : A(P') \to A(P)$ by $\mathcal{S}(G) : \mathcal{S}(N'_{\omega'}) \to \mathcal{S}(N_\omega)$. Then $A(g)$ is a homomorphism of groups; it is easy to see that A is a functor $\mathbf{P} \to \mathbf{A}$ which lifts Γ. This proves 4.11.

(4.12) THEOREM. *There is a multiplication making* Γ *into an* h-*space which is homotopy abelian and homotopy associative and which has a homotopy inversion. The resulting abelian group structure on* $[X,\Gamma]$ *coincides with that of Theorems* 4.4 *and* 4.10 *if* X *is homotopy equivalent to a polyhedron.*

An elementary proof of 4.12 is outlined in Section 6. Here we follow Lashof and Rothenberg [12] in using the following deep result.

(4.13). THEOREM. $[S^n,\Gamma]$ *is finite for all* $n \geq 0$.

Proof of 4.13. By 4.2 $[S^n,\Gamma]$ is isomorphic to the set of concordance classes of smoothings of the standard PL n-sphere S^n. From I, concordance implies orientation preserving diffeomorphism, while it is well known that the converse is true for S^n. For $n \geq 5$ and the generalized Poincaré conjecture of Smale [25], orientation preserving diffeomorphism is the same relation on smoothings of S^n as h-cobordism. Finiteness of the h-cobordism groups for $n \geq 5$ is proved by Kervaire and Milnor [12].

For $n \leq 3$ Munkres [22] and Smale [26] show that all smoothings of S^n are concordant, while Cerf [3] does the same for S^4. Thus 4.13 is proved.

(4.14) LEMMA. *Let* X *be a countable complex and* Y *a space with finite homotopy groups. Then the natural map*

$$\lambda : [X,\Gamma] \to \text{inv lim } [X_i,\Gamma]$$

is bijective, where X_i *denotes the* i*-skeleton of* X.

Proof. λ is always surjective. To prove λ injective, let $f : X \to \Gamma$ be such that for each i there exists a map $g_i : CX_i \to Y$ such that $g_i | X_i = f|X_i$. Here CX_i is the cone on X_i; it is assumed all the g_i take the vertex of the cone to y_0.

For each 0-cell a_j, $j = 1, 2, \ldots,$ of X, the maps $g_i|Ca_j$ fall into finitely many homotopy classes of paths with fixed end points. Starting with $j = 1$, we find a subsequence $\{g_{i_n}\}$, $n = 1, \ldots,$ of the g_i such that $g_{i_n}|Ca_j$, $n = 1, 2, \ldots,$ are all homotopic with fixed end points. Passing to $j = 2, 3, \ldots,$ we choose further subsequences until we find a subsequence, also denoted by $\{g_{i_n}\}$, such that for each j, $g_{i_n}|Ca_j \simeq g_{i_1}|Ca_j$, with fixed end points. Therefore $g_{i_n}|CX_0 \simeq g_{i_1}|CX_0$ rel X_0, all n. Since $i_n \geq n$, f_{i_n} is defined in CX_n. Let $h_n = CX_n \to Y$ be homotopic to g_{i_n} rel X_0, such that $h_n = g_{i_1}$ on CX_0. Thus we have found a single null homotopy of $f|X_0$ which extends to a null homotopy of $f|X_i$ for each X_i.

Proceeding inductively, suppose we have a map $g: CX_n \to Y$ which is a null homotopy of $f|X_n$, and which extends for each $i > n$ to a null homotopy $u_i : CX_i \to Y$ of $f|X_i$. An argument similar to the above shows that there is a subsequence $\{u_{i_m}\}$, $m = 1, 2, \ldots$, of the u_i, such that $u_{i_m}|CX_{n+1} \simeq u_{i_1}$ rel X_{n+1}. Define $g_{n+1} = u_{i_1}|CX_{n+1}$. Then $g_{n+1}|CX_n = g_n$, and g_{n+1} extends to a null homotopy of $f|X_j$ for each $j \geq n+2$. By induction we have found a sequence $g_n : CX_n \to Y$ such that $g_{n+1}|CX_n = g_n$, $g = 1, 2, \ldots$; and $g_n|X = f$. Define $g: CX \to Y$ by $g|CX_n = g_n$. Then g is a null homotopy of f.

(4.15) LEMMA. *Let* F *be the category of spaces having the homotopy type of countable* CW *complexes. There is a lifting of the functor* $\Gamma : F \to S$, $\Gamma(X) = [X, \Gamma]$, *into the category* A *of abelian groups, which extends the lifting of* $\Gamma : P \to S$ *of* 4.10. *Moreover if* $X \in F$, *then the natural map* $\lambda : [X, \Gamma] \to \text{inv lim } [X_i, X]$ *is an isomorphism of groups.*

Proof. From 4.14 we know that λ is an isomorphism of sets. The sequence $[X_0, \Gamma] \leftarrow [X_1, \Gamma] \leftarrow \cdots$ is an inductive sequence of homomorphisms of abelian groups. Hence the desired group structure on $[X, \Gamma]$ is $\text{ind} \lim_{i\to\infty} [X_i, \Gamma]$.

Proof of 4.12. Let F be as in 4.15. By 4.13 $\Gamma \times \Gamma$ is in F. Therefore there is an abelian group structure on $[\Gamma\times\Gamma, \Gamma]$. Define $\mu \in [\Gamma\times\Gamma, \Gamma]$ to be $[\pi_1] + [\pi_2]$, where $\pi_i : \Gamma \times \Gamma \to \Gamma$ is the projection on the ith factor. Then it is easy to see that μ makes Γ into an h-space, and hence $[X, \Gamma]$ has a natural abelian group structure for all spaces X, completing the proof of 4.12.

(4.16) THEOREM. *The fibration* $E = (P, BO, B\widetilde{PL})$ *is* Γ*-principal. If* $f: X \to B\,PL$ *is such that the fibration* f^*E *induced by* f *has a section,* *then* f^*E *is fibre homotopically trivial. Choosing a homotopy class of sections of* f^*E *induces a bijection between* $[X,\Gamma]$ *and homotopy classes of sections of* f^*E.

To say that E is Γ-principal means the following. There is a map $\mu : BO\times\Gamma \to BO$ such that the diagram

$$\begin{array}{ccc} BO \vee \Gamma & & \\ \downarrow & \searrow^{\mu_0} & \\ BO \times \Gamma & \xrightarrow{\ \mu\ } & BO \\ \downarrow & & \downarrow \\ BO & \xrightarrow{\ p\ } & B\,\widetilde{PL} \end{array}$$

commutes; where μ_0 is the natural map; and such that the two maps

$$(BO\times\Gamma)\times\Gamma \to BO\times\Gamma \to BO$$

$$BO\times(\Gamma\times\Gamma) \to BO\times\Gamma \to BO$$

are homotopic, the homotopy preserving the projections into $B\,\widetilde{PL}$. Moreover, it is required that on the fibre Γ of E over the base point, $\mu|\Gamma\times\Gamma$ is the h-space structure of Γ, and that $\mu|BO \vee \Gamma$ is the identity on BO and on Γ.

We first prove:

(4.17) LEMMA. *Let* X *and* Y *be countable* CW *complexes. Let*

$$\begin{array}{ccc} X \vee Y & \xrightarrow{\ f\ } & BO \\ \downarrow & & \downarrow p \\ X & \xrightarrow{\ g\ } & B\,PL \end{array}$$

be a commuting diagram. Then there is a commuting diagram

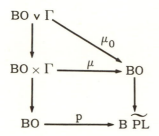

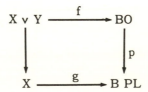

$$\begin{array}{ccc} X \times Y & \xrightarrow{F} & BO \\ \downarrow & & \downarrow \\ X & \xrightarrow{g} & B\,PL \end{array}$$

where F *extends* f.

Proof. We are assuming $X \vee Y$ is the union of X and Y with only $x_0 = y_0$ in common; that $p^{-1}g(x_0) = \Gamma$; and $f(x_0) =$ identity of Γ.

First suppose X and Y are finite polyhedrons.

Consider the composition

$$u : X \times Y \longrightarrow X \subset X \vee Y \xrightarrow{f} BO$$

as a lift of $p \circ u : X \times Y \to B\,PL$. Homotopy classes of lifts of $p \circ u$ correspond bijectively to the group $[X \times Y, \Gamma]$, with u corresponding to 0 by 4.12. Let $F : X \times Y \to BO$ be the lift of $p \circ u$ corresponding to the homotopy class of the composition

$$w : X \times Y \longrightarrow Y \subset X \vee Y \xrightarrow{f} \Gamma\ .$$

Note that $w|X \simeq 0$, while $w|Y = f|Y$. It follows that $F|X \vee Y$ is homotopic through lifts of $p \circ f$ to f. Therefore we may assume $F|X \vee Y = f$, proving 4.17 for finite complexes.

For countable complexes X, Y, let $F_i : X_i \times Y_i \to BO$ be the map just constructed. Then $F_{i+1}|X_i \times Y_i$ and F_i are homotopic lifts rel $X_i \vee Y_i$, by naturality of the correspondence between lifts and maps into Γ. Therefore each F_i extends over $X_{i+1} \times Y_{i+1}$; hence over $X_{i+2} \times Y_{i+2}$; etc. This proves 4.17. (Here $X_1 \subset X_2 \subset \ldots$ is an increasing sequence of finite complexes whose union is X; similarly for Y.)

In order to prove 4.16, observe that BO and Γ have the homotopy type of countable CW complexes. Therefore the map

$$(BO) \vee \Gamma \to BO$$

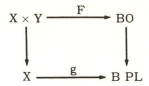

that is the identity on BO and on Γ, extends to a map $\mu: BO \times \Gamma \to BO$ making the diagram
$$\begin{array}{ccc} BO\times\Gamma & \to & BO \\ \downarrow & & \downarrow \\ BO & \to & BPL \end{array}$$
commute. Moreover, the construction of μ shows that $\mu|\Gamma \times \Gamma$ is the h-operation of Γ. For the last statement of 4.16, it follows from the construction that the two maps $(BO\times\Gamma\times\Gamma) \to BO$ are homotopic, as lifts of the compositions $BO \times \Gamma \times \Gamma \to BO \to B\,PL$, on every skeleton. Therefore, the map $BO \times \Gamma \times \Gamma \to \Gamma$, measuring the difference of lifts, is null homotopic on every skeleton. Therefore it is globally null homotopic by 4.13 and 4.14. This completes the proof of 4.16.

(4.18) COROLLARY. *Lemma* 4.17 *is true for arbitrary spaces* X, Y.

Proof. Let $f: X \vee Y \to BO$ be as in 4.17. Since $f(Y) \subset \Gamma$, we can factor f thus:

$$X \vee Y \xrightarrow{f_0} BO \vee \Gamma \longrightarrow BO\ ;$$

and f_0 extends to $f_1 . X \times Y \to BO$. Then $\mu \circ f_1 : X \times Y \to BO$ is the required extension f_1 where $\mu : BO \times \Gamma \to BO$ is from 4.16.

(4.19) COROLLARY. *Let* $E \to B$ *be a fibration induced from* $p: BO \to B\,PL$ *by a map* $f: B \to B\,PL$. *If* E *has a section then it is fibre homotopically trivial.*

Proof. The bundle E inherits a map $\mu': E\times\Gamma \to E$ from $\mu: BO\times\Gamma \to BO$. If $s: B \to E$ is a section, define $f: B\times\Gamma \to E$ to be the composition

$$B \times \Gamma \xrightarrow{s+1} E \times \Gamma \xrightarrow{\mu} E\ .$$

Then f is a fibre homotopy equivalence.

§5. *Obstruction theories*

The classification theorems 4.1 and 4.2 lead immediately to obstruction theories for *concordance classes* of smoothings. In order to obtain obstruction theories for *individual* smoothings the following *concordance extension theorem* is needed.

(5.1) THEOREM. *Let* M *be a* PL *manifold,* $A \subset M$ *a closed subset and* $U \subset M$ *an open neighborhood of* A. *Let* α_0 *be a smoothing of* M *and* α_1 *a smoothing of* $U \times I$ *which agrees with* α_0 *on* $U \times 0$. *Then there exists a smoothing* β *of* $M \times I$ *which agrees with* α_0 *on* $M \times 0$ *and with* α_1 *in a neighborhood of* $A \times I$.

Proof. The first step is to find a neighborhood $W \subset M \times I$ of $M \times 0 \cup A \times I = X$ and a smoothing γ of W agreeing with α_0 and α_1 in a neighborhood $V \subset M \times I$ of X. To do this, apply uniqueness of PD collarings (I, 13.10) to conclude that the germ of any PD collaring of part of the boundary of a smooth manifold can be extended to a collaring of the whole boundary. In the present situation this implies the existence of a PD embedding $f: N \to (U \times I)_{\alpha_1}$ where $N \subset U \times [0,1\rangle$ is an open neighborhood of $U \times 0$, having the following properties. There are closed sets B, C in M such that $A \subset \operatorname{int} B$, $B \subset \operatorname{int} C$, $C \subset U$, and

$$f = \text{identity on } N \cap [(U-C) \times I] \;;$$

and

$$f : N \cap (\operatorname{int} B \times I)_{\alpha_0 \times \iota} \to (U \times I)_{\alpha_1}$$

is a smooth embedding.

Let $W = ((M-C) \times I) \cup (\operatorname{int} B \times I) \cup N$. The smoothing γ of W is defined to be the product $\alpha_0 \times L$ on $(M-C) \times I$; α_1 on $N \cap (\operatorname{int} B \times I)$; and $f^* \alpha_1$ on N. The properties of f show that these three smoothings agree where they overlap; clearly γ has the desired properties.

The next step is to find a PL embedding $G: M \times I \to W$ which is the identity on $M \times 0$ and on a neighborhood $D \times I$ of $A \times I$; this is left to the reader.

Next use (I, 7.4) to push $G(M \times 1)$ onto a smooth submanifold of W_γ, by a PD embedding $H : G(M \times I) \to W$ which fixes $M \times 0$ and a neighborhood of $A \times I$. (The smoothing γ can be extended to a smoothing of $(M \times \langle -\infty, 0]) \cup W \cup \operatorname{int} B \times [1, \infty\rangle$ if the reader objects to the boundary of W in applying (I, 7.4).) Thus $H \circ G : M \times I \to W_\gamma$ is a PD embedding onto a smooth submanifold, leaving fixed $M \times 0$ and a neighborhood of $M \times I$. Therefore $(H \circ G)^* \gamma$ is the desired smoothing of $M \times I$.

An immediate consequence of the concordance extension theorem is:

(5.2) COROLLARY. *Let* A *be a closed subset of a* PL *manifold* M. *Let* α *be an* A-*germ of a smoothing of* M (*see* Section 2). *Then* α *extends to a smoothing of* M *if and only if there is a smoothing of* M *whose* A-*germ is concordant to* α.

Combining the classification theorem 4.2 and Corollary 5.2 we obtain:

(5.3) THEOREM. *Let* M *be a* PL *manifold and* $\mathcal{E}(M)$ *the fibration over* M *with fibre* $\widetilde{PL}/O$ *induced from* $B_O \to B_{\widetilde{PL}}$ *by a map* $M \to B_{\widetilde{PL}}$ *classifying the tangent block bundle of* M. *Let* $A \subset M$ *be a closed set and* α *an* A-*germ of a smoothing of* M. *Let* $s : A \to \mathcal{E}|A$ *be a section whose homotopy class corresponds to* α. *Then* α *extends to a smoothing of* M *if and only if* s *extends to a section of* $\mathcal{E}$.

From 5.3 we can interpret the obstructions to smoothing M as obstructions to a section of $\mathcal{E}$. These lie in groups

$$H^i(M; \Pi_{i-1}(\widetilde{PL}/O)) .$$

Conceivably the coefficients might be twisted, but actually this does not happen. To see why, observe that the fibre $\widetilde{PL}/O$ is connected because 1-manifolds are uniquely smoothable up to concordance. (In fact

$\widetilde{PL}/O$ is 6-connected.) Therefore E has a cross-section over the 1-skeleton of $B\,\widetilde{PL}$. On the other hand E is Γ-principal (4.16), hence E is fibre homotopically trivial over the 1-skeleton. This implies simple coefficients.

If $A \subset M$ is a closed set and α is a smoothing of a neighborhood of $M_{i-1} \cup A$ in M, the obstruction cohomology class to extending the $M_{i-2} \cup A$ germ of α over a neighborhood of $M_i \cup A$ lies in

$$H^i(M, A; \Pi_{i-1}(\widetilde{PL}/O)) .$$

It vanishes if and only if such an extension is possible.

Since a concordance is a smoothing, obstructions to extending a concordance are readily obtained. Let α_0, α_1 be smoothings of M and γ a concordance between their $M_{i-2} \cup A$ germs; γ is a smoothing of a neighborhood in $M \times I$ of $(M_{i-2} \cup A) \times I$. Assume γ extends to a concordance between the $M_{i-1} \cup A$ germs. The obstruction class to extending γ to a concordance between the $M_i \cup A$ germs of α_0 and α_1 lies in a group naturally isomorphic to

$$H^i(M, A; \Pi_i(\widetilde{PL}/O)) .$$

It vanishes if and only if such an extension of γ is possible.

Obstructions to smoothing a PD homeomorphism are obtained as follows. Let M be a triangulated PL manifold with smoothing α. Let N_β be a smooth manifold and $f : M \to N_\beta$ a PD homeomorphism. Suppose f is a diffeomorphism in a neighborhood of $M_{i-1} \cup A$. The problem is to find a PD isotopy f_t of $f = f_0$ so that f_1 is a diffeomorphism in a neighborhood of $M_i \cup A$; it is required that $f_t = f$ in a neighborhood of $M_{i-2} \cup A$. The smoothings α and $f^*\beta$ agree in a neighborhood of $M_{i-1} \cup A$, and the desired isotopy exists if and only if α and $f^*\beta$ are concordant in a neighborhood of $M_i \cup A$, the concordance being a product in a neighborhood of $(M_{i-2} \cup A) \times I$. Therefore the obstruction to finding such an isotopy is defined as the obstruction to finding such a concordance.

It lies in

$$H^i(M, A; \Pi_i(\widetilde{PL}/O)) ,$$

and vanishes if and only if there is a PD isotopy $f_t : M \to N$ as above.

Munkres [20, 21] shows that these obstructions are dual to his homology obstructions [18, 19].

Another interpretation of the problem of smoothing f is as follows. $f^*\beta$ is concordant to $f^*\alpha$ if and only if the two linearizations $\mathrm{Exp}(\alpha)$ and $\mathrm{Exp}(f^*\beta)$ of $(M \times M, M_\Delta)$ are equivalent (3.10). Let $t(\alpha)$, $t(\beta)$ denote respectively the tangent PD vector bundles of M_α, N_β, considered as linearizations of $(M \times M, M_\Delta)$ and $(N \times N, N_\Delta)$. The map $f \times f : M \times M \to N \times N$ restricts to what might be called a PD map $\delta f : t(\alpha) \to t(\beta)$, covering $f : M \to N$. Then f *is* PD *isotopic to a diffeomorphism* $M_\alpha \to N_\beta$ *if and only if* δf *is* PD *isotopic to a vector bundle isomorphism* $t(\alpha) \to t(\beta)$. Perhaps δf should be called the "combinatorial" of f, in analogy with the "differential" of a differentiable map.

§6. h-*structures on* BO, B PL *and* Γ

In this section we give a new description of the h-structure on Γ. Details are omitted since Boardman and Vogt [1] promise a thorough treatment of the whole subject.

We shall work in the semisimplicial category.

PL(n) is the semisimplicial group whose k-simplices are PL automorphisms (preserving 0-sections) of the trivial R^n bundle over Δ_k; $PL = \lim_{n\to\infty} PL(n)$. B PL(n) is a classifying space for PL(n), and $B\,PL = \lim_{n\to\infty} B\,PL(n)$ is then a classifying space for PL.

The stable tubular neighborhood theorem for the PL category [8, 13] implies that every block bundle is stably isomorphic to a PL bundle, the latter being unique up to stable isomorphism. This means we can identify $B\,\widetilde{PL}$ and B PL. Henceforth we ignore $\widetilde{PL}$.

(6.1) *The map* $t_n : BO(n) \to B\,PL(n)$

We follow Lashof and Rothenberg in defining t_n.

We replace the orthogonal group $\hat{O}(n)$ by the semisimplicial group complex $O(n)$ of PD singular simplices of $\hat{O}(n)$. Lashof and Rothenberg define a choice of $BO(n)$ and a map $BO(n) \to B\,PL(n)$ as follows. Let $p(n): E\,PL(n) \to B\,PL(n)$ be a universal principal bundle for $PL(n)$. Thus $PL(n)$ acts freely on $E\,PL(n)$ on the right with orbit space $B\,PL(n)$, and $E\,PL(n)$ is contractible. Since the inclusion $PD(n) \to PL(n)$ is a homotopy equivalence, the associated bundle

$$EO(n) = E\,PL(n) \times_{PL(n)} PD(n)$$

is also contractible. The right action of $O(n)$ on $PD(n)$ induces a free action of $O(n)$ on $EO(n)$. The orbit space of this action is then

$$BO(n) = E\,PL(n) \times_{PL(n)} (PD(n)/O(n)) \ .$$

The projection

$$E\,PL(n) \times PD(n) \to E\,PL(n)$$

induces a semi-simplicial fibration

$$PD(n)/O(n) \subset BO(n) \xrightarrow{t_n} B\,PL(n) \ ,$$

and in the limit, another fibration

(1) $$PD/O \subset BO \xrightarrow{t} B\,PL \ .$$

The multiplications on BO and $B\,PL$ induce the operation of Whitney sums on stable bundles.

Let $\Lambda : \mathcal{R}^\infty \times \mathcal{R}^\infty \to \mathcal{R}^\infty$ be an isometry (= norm preserving bijective linear map). Define homomorphisms

$$\lambda^O : O \times O \to O \ , \quad \lambda^{PL} : PL \times PL \to PL$$

by

$$(f, g) \mapsto \lambda(f \times g)\lambda^{-1} \ .$$

The induced maps on classifying spaces are denoted

$$\mu_O : BO \times BO \to BO\,, \quad \mu_{PL} : B\,PL \times B\,PL \to B\,PL\ .$$

(6.2) THEOREM. *μ_O and μ_{PL} are homotopy abelian, homotopy associative h-space structures. Moreover μ_O can be chosen within its homotopy class so that the following diagram commutes*:

$$\begin{array}{ccc} BO \times BO & \xrightarrow{\mu_O} & BO \\ \downarrow{\scriptstyle t \times t} & & \downarrow{\scriptstyle t} \\ B\,PL \times B\,PL & \xrightarrow{\mu_{PL}} & B\,PL \end{array}$$

where t *is the fibration* (1). *Consequently the fibre* Γ *of* t *is an h-space.* Γ *is also homotopy associative and homotopy commutative.*

Proof. For $0 < k < q$ let ${}_qO(k) \subset O(q)$ be the semisimplicial subgroup of PD singular simplices whose values leave fixed the first $q-k$ coordinates. Put

$$O(q, n) = O(q)/{}_qO(q-n)\ .$$

This is the same as the PD singular complex of the Stiefel manifold $V_{q,n}$ of orthogonal n-frames in $\mathcal{R}^q$.

Put $PL(q, n) = PL(q)/{}_qPL(q-n)$. Define

$$O(\infty, n) = \text{dir lim}_{q\to\infty}\, O(q, n)$$

$$O(\infty, \infty) = \text{inv lim}_{n\to\infty}\, O(\infty, n)\ .$$

Then

$$\pi_i(O(\infty, n)) = \text{dir lim}_{q\to\infty}\, \pi_i(V_{q,n}) = 0$$

so $O(\infty, n)$ is contractible for each n. Since each map $O(\infty, n+1) \to O(\infty, n)$ is a fibration, $O(\infty, \infty)$ is contractible.

Let $\text{Hom}(O(n), O(q))$ be the PD singular complex of $\text{Hom}(\hat{O}(n), \hat{O}(q))$, the space of homomorphisms $\hat{O}(n) \to \hat{O}(q)$. Define a map

$$\Phi^{O}_{q,n} : O(q, n) \quad \mathrm{Hom}\,(O(n), O(q))$$

as follows. If a singular simplex $g : \Delta_k \to \hat{O}(q)$ represents the coset $[g] \in O(q)/_q O(q-n) = O(q, n)$ then $\Phi^{O}_{q,n}[g]$ is the singular simplex $\Delta^k \to \mathrm{Hom}\,(\hat{O}(n), \hat{O}(q))$ whose value at $x \in \Delta^k$ is the homomorphism

$$h \mapsto g(x)\,(h \times 1)\,g(x)^{-1} \ .$$

Here $h \in \hat{O}(n)$, and 1 denotes the identity map of $\mathcal{R}^{q-n}$; we identify $\mathcal{R}^n \times \mathcal{R}^{q-n}$ with $\mathcal{R}^q$. It is easily verified that $\Phi^{O}_{q,n}$ is well defined.

Let $\mathrm{Hom}\,(O(n), O) = \mathrm{dir}\,\lim_{q\to\infty} \mathrm{Hom}\,(O(n), O(q))$ and $\mathrm{Hom}\,(O, O) = \mathrm{inv}\,\lim_{n\to\infty} \mathrm{Hom}\,(O(n), O)$. Define $\Phi^{O}_{n} : O(\infty, n) \to \mathrm{Hom}\,(O(n), O)$ to be the direct limit of the maps $\Phi^{O}_{q,n}$; define $\Phi^{O} : O(\infty, \infty) \to \mathrm{Hom}\,(O, O)$ to be the inverse limit of the maps Φ^{O}_{n}.

A semisimplicial complex $\mathrm{Hom}\,(PL(n), PL(q))$ is defined as follows. Let $\mathbf{\Delta}^k$ denote the semisimplicial complex of non-decreasing finite sequences of vertices of Δ^k. A k-simplex of $\mathrm{Hom}\,(PL(n), PL(q))$ is a semisimplicial map

$$f : \mathbf{\Delta}^k \times PL(n) \to PL(q)$$

such that for each r-simplex σ of Δ^k, $(q = 0, 1, \ldots)$ the map

$$[PL(n)]_r \to [PL(q)]_r \ ,$$

given by $x \mapsto f(\sigma, x)$, is a homomorphism. Taking direct and then inverse limits defines complexes $\mathrm{Hom}\,(PL(n), PL)$ and $\mathrm{Hom}\,(PL, PL)$.

Define $\Phi^{PL}_{q,n} : PL(q, n) \to \mathrm{Hom}\,(PL(n), PL(q))$ as follows. Let $q \in PL(q)_r$ represent $[g] \in [PL(q, n)]_r$. Then $\Phi^{PL}_{q,n}[g] \in [\mathrm{Hom}(PL(n), PL(q))]_r$ is the unique semisimplicial map $\mathbf{\Delta}_r \times PL(n) \to PL(q)$ which assigns to (δ_r, h) the PL automorphism

$$g(h \times 1)\,g^{-1}$$

of the trivial bundle $\Delta_r \times \mathcal{R}^q$. Here δ_r is the unique nondegenerate r-simplex of $\mathbf{\Delta}_r$. By taking direct and then inverse limits, define maps

$$\Phi_n^{PL} : PL(\infty, n) \to Hom(PL(n), PL)$$

and

$$\Phi^{PL} : PL(\infty, \infty) \to Hom(PL, PL) \ .$$

Let $\lambda' : R^\infty \to R^\infty$ be the composition

$$\lambda' : \mathcal{R}^\infty \to \mathcal{R}^\infty \times 0 \subset \mathcal{R}^\infty \times \mathcal{R}^\infty \xrightarrow{\lambda} \mathcal{R}^\infty \ ,$$

where λ is the isometry chosen previously. We consider λ' as a vertex of $O(\infty, \infty)$: for each integer $n \geq 0$ let $\nu(n)$ be the smallest integer m such that $\lambda'(\mathcal{R}^n) \subset \mathcal{R}^m$. Then $\lambda'|\mathcal{R}^n$ is a vertex of $O(n, \nu(n))$, that is, an orthogonal linear map $\mathcal{R}^n \to \mathcal{R}^{\nu(n)}$; hence $\lambda'|\mathcal{R}^n$ is a vertex of $O(n,\infty)$. The sequence $\lambda'|\mathcal{R}^n$, $n = 0, 1, \ldots,$ defines an element of $O(\infty, \infty)$.

Let $\alpha \in O(\infty, \infty)$ be the vertex defined by λ'.

It is easy to see that $\Phi^O(\alpha) \in Hom(O, O)$ is the homomorphism $O \to O$ which sends $g \in O$ to $\lambda^O(g \times 1)$, where $\lambda^O : O \times O \to O$ is the homomorphism defined above. Similarly, $\Phi^{PL}(\alpha) \in Hom(PL, PL)$ is the homomorphism sending $g \in PL$ to $\lambda^{PL}(g \times 1)$.

Every homomorphism $f : G \to H$ between semisimplicial groups includes a map $B(f) : BG \to BH$ as follows. Since the universal principal H bundle EH is contractible there is a unique homotopy class of sections of the bundle over BG associated to EG with fibre EH, letting G act on EH via f. Such a section is the same thing as a G-equivariant map $EG \to EH$. Such a map covers the map $B(f) : BG \to BH$.

If f_0 and f_1 are homotopic through homomorphisms then $B(f_0)$ and $B(f_1)$ are homotopic. This is proved by considering equivariant maps $(EG) \times I \to (EH) \times I$.

We now prove that $\mu : BO \times BO \to BO$ is an h-structure. We must show that the composite map

$$\mu' : BO \longrightarrow BO \times \{x_0\} \longrightarrow BO \times BO \xrightarrow{\mu} BO$$

is homotopic to the identity, where x_0 is the base point of BO. Now it is easy to see that $\mu' = B(\Phi^O(\alpha))$. Since $O(\infty, \infty)$ is contractible, α is connected to the identity. Clearly $\Phi^O(\varepsilon)$ is the identity $\varepsilon \in O(\infty,\infty)$, where ε is the inverse limit of the inclusions $O(n) \to O(\infty)$. Hence $\mu' = B(\Phi^O(\varepsilon))$. Clearly $\Phi^O(\varepsilon)$ is the identity homomorphism of O; so μ' is homotopic to the identity. Likewise, the composition

$$\mu' : BO \longrightarrow \{x_0\} \times BO \longrightarrow BO \times BO \xrightarrow{\mu} BO$$

is homotopic to the identity.

In a similar way, using the connectedness of $O_{\infty,\infty}$, it is proved that μ is homotopy associative and homotopy commutative.

A similar argument can be used for $B\,PL$. Instead of proving $PL(\infty,\infty)$ contractible, we construct a map $O(\infty,\infty) \to PL(\infty,\infty)$ which extends the natural map of vertices. Such a map proves that $PL(\infty,\infty)$ is connected, and then the same proof shows that $B\,PL$ is a homotopy commutative and homotopy associative h-space.

We must define maps $O(q,n) \to PL(q,n)$ which extend the natural map of vertices; which commute with the inclusions

$$O(q,n) \to O(q,n+1)$$

and

$$PL(q,n) \to PL(q,n+1) \;;$$

and which also commute with the fibrations

$$O(q,n+1) \to O(q,n)$$

and

$$PL(q,n+1) \to PL(q,n) \;.$$

We start with the Lashof-Rothenberg map $t_q : BO(q) \to B\,PL(q)$ of 6.1 above. A similar construction produces a map $B({}_qO(q-n)) \to B({}_qPL(q-n))$, and the following diagram commutes up to homotopy:

$$\begin{array}{ccc} B({}_qO(q-n)) & \longrightarrow & BO(q) \\ \downarrow & & \downarrow \\ B({}_qPL(q-n)) & \longrightarrow & B\,PL(q) \;. \end{array}$$

Each diagram

$$
(6.3) \qquad \begin{array}{ccc} O(q,n) & \subset & O(q+1,n) \\ \downarrow & & \downarrow \\ PL(q,n) & \subset & PL(q+1,n) \end{array}
$$

commutes up to homotopy, since with proper choices of universal bundles it commutes exactly. By recursion on q, using the homotopy extension property for the pairs $(O(q+1,n), O(q,n))$ we adjust each vertical map within its homotopy class so that (6.3) commutes for all (q,n). Therefore a map $O(\infty,n) \to PL(\infty,n)$ is obtained.

Each diagram

$$
(6.4) \qquad \begin{array}{ccc} O(\infty,n) & \longleftarrow & O(\infty,n+1) \\ \downarrow & & \downarrow \\ PL(\ \ ,n) & \longleftarrow & PL(\ \ ,n+1) \end{array}
$$

commutes up to homotopy, and the vertical maps are fibrations. By recursion on n we adjust each vertical map in succession until 6.4 commutes exactly for all n. The result is the desired map $O(\infty,\infty) \to PL(\infty,\infty)$.

This completes the proof that BO and $B\,PL$ are homotopy abelian h-groups. It follows that the fibre Γ of $t: BO \to B\cdot PL$ is an h-group. To know that Γ is homotopy-commutative and homotopy associative, we need to show that $t: BO \to B\,PL$ commutes with homotopies of associativity and commutativity. This is true and is a consequence of the constructions used. This completes the proof of 6.2.

It is not hard to see that the multiplications on BO and $B\,PL$ are compatible with Whitney sum maps: $BO(n)\times BO(m) \to BO(n+m)$ and $B\,PL(n)\times B\,PL(m) \to B\,PL(n+m)$.

It is left to the reader to verify that for finite dimensional CW complexes, X, the group structure on $[X,\Gamma]$ is the same using either the multiplication on Γ defined in this section or the preceding one.

Remarks:

1. $PL(\infty, n)$ and $PL(\infty, \infty)$ are contractible.
2. It seems likely that by exploiting the contractibility of $PL(\infty, \infty)$ and $O(\infty, \infty)$ (and not just their connectedness) one could show directly that the h-spaces Γ, BO and B PL satisfy the conditions A(n) of Stasheff [27] for all n, which would mean they have classifying spaces. There should also be maps $B\Gamma \to B(BO) \to B(B\,PL)$. Boardman and Vogt claim that Γ and B PL are in fact infinite loop spaces (as is known for BO), and hence have classifying spaces.
3. It is well known that BO is a group. Perhaps the shortest construction is to observe that O is a normal subgroup of the contactible group O_∞ of isometries of $\mathfrak{R}^\infty$. Hence O_∞/O is a group which can be used for BO.

BIBLIOGRAPHY

[1]. Boardman, J., and Vogt, R., Homotopy everything H-spaces. Bull. Amer. Math. Soc. 74 (1968), 1117-1122.

[2]. Cairns, S., Homeomorphisms between topological manifolds and analytic manifolds. Annals Math. 41 (1940), 796-808.

[3]. Cerf, J., Sur les difféomorphismes de la sphère de dimension trois ($\Gamma_4 = 0$). Lecture notes in Mathematics. 53 (1968), Springer-Verlag.

[4]. Haefliger, A., Plongement différentiables de variétés dans variétés. Comm. Math. Helv. 36 (1961), 48-82.

[5]. ————, Knotted spheres and related geometrical problems. Proc. Int. Congress of Math., Moscow, 1966.

[6]. Hirsch, M., Obstructions to smoothing manifolds and maps. Bull. Amer. Math. Soc. 69 (1963), 352-356.

[7]. ————, On tubular neighborhoods of piecewise linear and topological manifolds, *Proc. of Conference on Manifolds*, Prindle, Weber and Schmidt 1968, 63-80.

[8]. ————, On normal microbundles. Topology 5 (1966), 229-240.

[9]. Hudson, J., Extending piecewise linear isotopies. Proc. London Math. Soc. (3), 16 (1966), 651-668.

[10]. ————, Piecewise linear embeddings and isotopies. Bull. Amer. Math. Soc. 72 (1966), 536-7.

[11]. Kato, M., Combinatorial prebundles I, II. Osaka J. Math. 4 (1968).

[12]. Kervaire, M., and Milnor, J., Groups of homotopy spheres, part I. Annals Math. 77 (1963), 504-577.

[13]. Lashof, R., and Rothenberg, M., Microbundles and smoothing. Topology 3 (1965), 357-380.

[14]. Mazur, B., and Poenaru, V., Seminaire de Topologie combinatoire et differentielle. Inst. Hautes Etudes Scient. 1962.

[15]. Milnor, J., Microbundles, Part I. Topology 3, supplement 1 (1964), 53-80.

[16]. ________, Topological manifolds and smooth manifolds. Proc. Int. Cong. Math., Stockholm 1962, 132-138.

[17]. Morlet, C., Voisinages tubulaires des variétés semilineaires. C. R. Acad. Sci. Paris, 262 (1966), 740-743.

[18]. Munkres, J., Obstructions to the smoothing of piecewise differentiable homeomorphisms. Annals Math. 72 (1960), 521-554.

[19]. ________, Obstruction to imposing differentiable structures. Ill. J. Math. 8 (1964), 361-376.

[20]. ________, Concordance is equivalent to smoothability. Topology 5 (1966), 371-389.

[21]. ________, Compatibility of imposed differentiable structures. Ill. J. Math. 12 (1968), 610-615.

[22]. ________, Differentiable isotopies on the 2-sphere. Mich. Math. J. 7 (1960), 193-197.

[23]. Rourke, C., and Sanderson, B., Block bundles. Ann. Math. 87 (1968), 1-28; 256-278; 431-483.

[24]. ________________, Δ-sets. University of Warwick 1969 (mimeographed).

[25]. Smale, S., Generalized Poincaré's conjecture in dimensions greater than 4. Annals Math. 74 (1961), 391-406.

[26]. ________, Diffeomorphisms of the 2-sphere. Proc. Amer. Math. Soc. 10 (1959), 621-626.

[27]. Stasheff, J., Homotopy associativity of H-spaces I. Trans. Amer. Math. Soc. 108 (1963), 275-292.

[28]. Steenrod, N., *The Topology of Fibre Bundles*. Princeton, 1951.

[29]. Thom, R., Des variétés triangulées aux variétés différentiables. Proc. Int. Cong. Math., Edinburgh 1958, 248-255.

[30]. White, D., Smoothing of embeddings and classifying spaces. Thesis, U. Geneva, 1971.

Library of Congress Cataloging in Publication Data

Hirsch, Morris, 1933-
Smoothings of piecewise linear manifolds.

(Annals of mathematics studies, no. 80)
Bibliography: p.
1. Piecewise linear topology. 2. Manifolds (Mathematics) I. Mazur, Barry, joint author. II. Title. III. Series.
QA613.4.H57 514'.224 74-2967
ISBN 0-691-08145-X

ANNALS OF MATHEMATICS STUDIES

Edited by Wu-chung Hsiang, John Milnor, and Elias M. Stein

79. Discontinuous Groups and Riemann Surfaces, *edited by* LEON GREENBERG
78. Strong Rigidity is Locally Symmetric Spaces *by* G. D. MOSTOW
77. Stable and Random Motions in Dynamical Systems: With Special Emphasis on Celestial Mechanics, *by* JÜRGEN MOSER
76. Characteristic Classes, *by* JOHN W. MILNOR and JAMES D. STASHEFF.
75. The Neumann Problem for the Cauchy-Riemann Complex, *by* G. B. FOLLAND *and* J. J. KOHN
74. Lectures on p-Adic L-Functions *by* KENKICHI IWASAWA
73. Lie Equations *by* ANTONIO KUMPERA *and* DONALD SPENCER
72. Introduction to Algebraic K-Theory, *by* JOHN MILNOR
71. Normal Two-Dimensional Singularities, *by* HENRY B. LAUFER
70. Prospects in Mathematics, *by* F. HIRZEBRUCH, LARS HÖRMANDER, JOHN MILNOR, JEAN-PIERRE SERRE, *and* I. M. SINGER
69. Symposium on Infinite Dimensional Topology, *edited by* R. D. ANDERSON
68. On Group-Theoretic Decision Problems and Their Classification, by CHARLES F. MILLER, III
67. Profinite Groups, Arithmetic, and Geometry, *by* STEPHEN S. SHATZ
66. Advances in the Theory of Riemann Surfaces, *edited by* L. V. AHLFORS, L. BERS, H. M. FARKAS, R. C. GUNNING, I. KRA, *and* H. E. RAUCH
65. Lectures on Boundary Theory for Markov Chains, *by* KAI LAI CHUNG
64. The Equidistribution Theory of Holomorphic Curves, *by* HUNG-HSI WU
63. Topics in Harmonic Analysis Related to the Littlewood-Paley Theory, *by* ELIAS M. STEIN
62. Generalized Feynman Amplitudes, *by* E. R. SPEER
61. Singular Points of Complex Hypersurfaces, *by* JOHN MILNOR
60. Topology Seminar, Wisconsin, 1965, *edited by* R. H. BING *and* R. J. BEAN
59. Lectures on Curves on an Algebraic Surface, *by* DAVID MUMFORD
58. Continuous Model Theory, *by* C. C. CHANG *and* H. J. KEISLER
57. Seminar on the Atiyah-Singer Index Theorem, *by* R. S. PALAIS
56. Knot Groups, *by* L. P. NEUWIRTH
55. Degrees of Unsolvability, *by* G. E. SACKS
54. Elementary Differential Topology (Rev. edn., 1966), *by* J. R. MUNKRES
52. Advances in Game Theory, *edited by* M. DRESHER, L. SHAPLEY, *and* A. W. TUCKER
51. Morse Theory, *by* J. W. MILNOR
50. Cohomology Operations, *lectures by* N. E. STEENROD, *written and revised by* D.B.A. EPSTEIN
48. Lectures on Modular Forms, *by* R. C. GUNNING
47. Theory of Formal Systems, *by* R. SMULLYAN
46. Seminar on Transformation Groups, *by* A. BOREL *et al.*
44. Stationary Processes and Prediction Theory, *by* H. FURSTENBERG
43. Ramification Theoretic Methods in Algebraic Geometry, *by* S. ABHYANKAR

PRINCETON UNIVERSITY PRESS

PRINCETON, NEW JERSEY 08540